Foundations of Physical Chemistry: Worked Examples

Nathan Lawrence

St. John's College, University of Oxford

Jay Wadhawan

St. John's College, University of Oxford

Richard Compton

Physical and Theoretical Chemistry Laboratory and St. John's College, University of Oxford

Series sponsor: **ZENECA**

ZENECA is a major international company active in four main areas of business: Pharmaceuticals, Agrochemicals and Seeds, Speciality Chemicals, and Biological Products.

ZENECA's skill and innovative ideas in organic chemistry and bioscience create products and services which improve the world's health, nutrition, environment, and quality of life.

ZENECA is committed to the support of education in chemistry and chemical engineering.

OXFORD
UNIVERSITY PRESS

Great Clarendon Street, Oxford OX2 6DP

Oxford University Press is a department of the University of Oxford
and furthers the University's aim of excellence in research, scholarship,
and education by publishing worldwide in

Oxford New York
Athens Auckland Bangkok Bogotá Buenos Aires Calcutta
Cape Town Chennai Dar es Salaam Delhi Florence Hong Kong Istanbul
Karachi Kuala Lumpur Madrid Melbourne Mexico City Mumbai
Nairobi Paris São Paulo Singapore Taipei Tokyo Toronto Warsaw

and associated companies in Berlin Ibadan

Oxford is a registered trade mark of Oxford University Press

Published in the United States
by Oxford University Press Inc., New York

British Library Cataloguing in Publication Data

Data available

Library of Congress Cataloging in Publication Data

Lawrence, Nathan.
Foundations of physical chemistry: worked examples / Nathan
Lawrence, Jay Wadhawan, Richard Compton.
(Oxford chemistry primers; 68)
Includes index.
1. Chemistry, Physical and theoretical. I. Wadhawan, Jay.
II. Compton, R.G. III. Title. IV. Series.
QD453.2.L38 1999 541.3—dc21 98–31454

ISBN 0-19-850462-4 (Pbk)

1 3 5 7 9 10 8 6 4 2

Typeset by EXPO Holdings, Malaysia

Printed in Great Britain
on acid-free paper by
The Bath Press, Avon

Founding Editor's Foreword

Foundations of physical chemistry: worked examples is the companion Primer to the internationally highly acclaimed *Foundations of physical chemistry* (OCP 40). Together these Primers are designed to facilitate the transition from chemistry studies at School to those at University by providing a concise and elementary coverage of core topics both descriptively and by way of worked examples.

Foundations of physical chemistry: worked examples has been extremely well constructed in a very user friendly format by a master chemist and two of his newly qualified apprentices. Students who aspire to take up apprenticeship in chemistry as well as their teachers will find this Primer invaluable.

Stephen G. Davies
The Dyson Perrins Laboratory, University of Oxford

Preface

This Oxford Chemistry Primer and the sibling text *Foundations of physical chemistry* (OCP 40) seek to develop a foundation in physical chemistry suitable for students commencing university courses in chemistry and allied subjects. It aims to link some material familiar from school studies (Advanced Level) with a selection of essential ideas usually encountered early in the freshman year. The first two chapters consider the structure of atoms and molecules. Chapters 3, 4 and 5 discuss chemical reactivity through energetics, kinetics and equilibria. Chapter 6 gives a brief insight into a few more advanced areas. We hope that the book will be accessible and stimulating to those studying Advanced Level chemistry whilst providing a sound basis for university work. Some more challenging material is additionally included for which, while in the bulk of the text the mathematical aspects of the subject have been minimized as far as is consistent with clarity, an elementary knowledge of calculus is utilised.

We thank all those who have so generously given us much valuable constant criticism and advice. We also thank those who suggested problems from their own spheres of teaching: Don Bethell (Liverpool University), Robert Dryfe (UMIST), Julie Macpherson (Warwick University) and Jason Riley (Bristol University). We also express our appreciation to those who have helped in diverse other ways not least in the wordprocessing aspects: Frank Marken, Jon Ball, Marco Fidel-Suárez, Kelvin Lawrence and Rita Wadhawan. However, above all we thank John Freeman who constructed all the diagrams in this primer. His artistic talents and patience are hugely admired.

July 23, 1998
Oxford

N. S. L.
J. D. W.
R. G. C.

Contents

Values of fundamental constants

Quantity	**Symbol**	**Value and units**
Speed of light	c	2.998×10^{8} m s^{-1}
Planck's constant	h	6.626×10^{-34} J s
Mass of an electron	m_e	9.109×10^{-31} kg
Atomic mass unit	u	1.661×10^{-27} kg
Elementary charge	e	1.602×10^{-19} C
Vacuum permittivity	ε_o	8.854×10^{-12} J C^{2} m^{-1}
Avogadro constant	N_A	6.022×10^{23} mol^{-1}
Faraday constant	F	9.6485×10^{4} C mol^{-1}
Universal gas constant	R	8.3145 J K^{-1} mol^{-1}

1 Atoms and ions: the building blocks of matter

1.1 Aims

This chapter will present questions, and their answers, aimed at illustrating various topics central to the study of chemistry as follows.

- Atomic size and the Avogadro constant.
- The structure of the hydrogen atom and its electronic spectrum.
- The structure of many-electron atoms.
- The Pauli Exclusion principle.

1.2 Worked examples

Question 1

(a) (i) Metallic lithium adopts a body-centred cubic (bcc) structure, as shown in Fig. 1.1(a). If the closest distance of approach of the nuclei is 3.03 Å and the density of the metal is 0.534 g cm^{-3}, estimate the Avogadro constant (N_A). The relative atomic mass of lithium is 6.94.

1 Å $= 10^{-10}$ m.

(ii) The bond length of the diatomic molecule Li_2 is 2.67 Å. Comment.

(b) In metallic lithium each atom is surrounded by fourteen near neighbours, eight at the distance of closest approach, 3.03 Å, and six more at 3.50 Å. Copper adopts a face-centred cubic (fcc) structure whilst that of zinc is hexagonal close packed (hcp) (Fig. 1.1(b) and (c)). How many nearest neighbours are there to any atom in each of the fcc and hcp structures?

(c) In the bcc structure, what percentage of the volume is occupied?

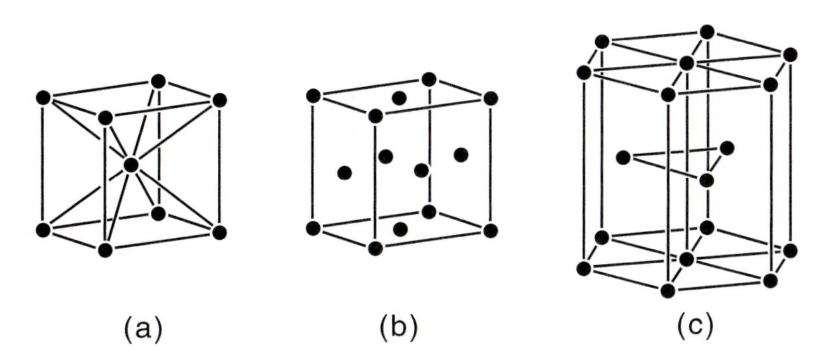

Fig. 1.1 Typical metallic structures. (a) Body-centred cubic (bcc); (b) face-centred cubic (fcc); (c) hexagonal close-packed (hcp).

One mole of an ideal gas occupies 22414 cm^3 at 0 °C and one atmosphere pressure. Some volumes (cm^3) occupied by real gases are He, 22396; N_2, 22403; Ar, 22390; Cl_2, 22063. The deviations reflect the presence of weak, but attractive intermolecular forces in real as compared to ideal gases. Strictly speaking, real gases therefore only follow the ideal gas law, $PV = nRT$ (where P is the gas pressure; V, its volume; n, the number of moles of ideal gas; R, the universal gas constant of 8.31 J K^{-1} mol^{-1}; and T, the absolute temperature) in the limit of zero pressure compressing the gas. Nevertheless, the closeness of the molar volumes given above to the ideal value testifies to the excellent nature of the ideal gas approximation.

(d) The radius of a helium atom is estimated as 1.2 Å. One mole of helium gas occupies 22396 cm^3 at 0 °C and one atmosphere pressure. What percentage of the volume is occupied?

(e) When radium atoms decay they emit α-particles which, when suitably collected, pick up electrons to form neutral helium atoms. If the rate of emission is ca. 11.6×10^{17} per gramme of radium in one year, estimate the Avogadro constant, if 0.043 cm^3 (as measured at 0 °C and one atmosphere pressure) of helium are formed in that time.

Answer

(a) (i) The unit cell of lithium shown in Fig. 1.1(a) contains a total of two lithium atoms: one atom is in the centre of the cell and the other is made up of eight contributions of $\frac{1}{8}$th of an atom from the atoms at the corners of the unit cell. The latter contribute $\frac{1}{8}$th since they are shared between eight unit cells in the crystal lattice.
The distance of closest approach of lithium atoms is 3.03 Å. This corresponds to the distance between the central lithium atom in Fig. 1.1(a) and a corner of the unit cell cube. If the length of one edge of the cube is d then,

$$\frac{\sqrt{3}}{2}\mathrm{d} = 3.03\ \text{Å}$$

so that

$$\mathrm{d} = 3.49\ \text{Å}.$$

The volume of the cube is therefore

$$\text{volume} = (3.49 \times 10^{-8})^3\ \mathrm{cm}^3.$$

It follows that,

$$\text{density} = \frac{\text{mass}}{\text{volume}} = \frac{2 \times 6.94}{(3.49 \times 10^{-8})^3 \times N_A} = 0.534\ \mathrm{g\ cm^{-3}}$$

where N_A is the Avogadro constant. Solving gives

$$N_A = 6.1 \times 10^{23}\ \mathrm{mol^{-1}}.$$

(ii) The stated lithium to lithium distance in the metal (3.03 Å) is significantly longer than in the diatomic molecule (Li_2) implying that the bonds in the metal are weaker. There are however more of them so that the binding energy of the metal actually exceeds that of the molecule.

(b) In both the fcc and hcp structures there are twelve equidistant neighbours.

(c) The bcc structure in Fig. 1.1(a) contains a total of two atoms each of radius r. The "filled" volume is therefore

$$2 \times \frac{4}{3}\pi r^3.$$

Assuming the centre atom in the unit cell 'touches' those in the corners, the length of one side of the unit cell is

$$\frac{4r}{\sqrt{3}} = 2.31r,$$

so that the volume of the unit cell is $(2.31r)^3$.

The percentage of the volume occupied is therefore

$$\frac{2 \times 4\pi}{3 \times 2.31^3} \times 100 = 68\%.$$

(d) The volume of a helium atom is

$$\frac{4}{3}\pi r^3 = \frac{4}{3}\pi(1.2 \times 10^{-8})^3 = 7.2 \times 10^{-24}\ \text{cm}^3.$$

In one mole of helium there is an Avogadro constant of molecules which therefore occupy a total volume of

$$6.1 \times 10^{23} \times 7.2 \times 10^{-24} = 4.4\ \text{cm}^3.$$

These are distributed in 22396 cm^3 so that the percentage of volume occupied is

$$\frac{4.4}{22396} \times 100 = 0.02\%.$$

(e) A volume of 0.043 cm^3 corresponds to the following number of moles of helium,

$$\frac{0.043}{22396} = 1.9 \times 10^{-6}\ \text{moles}.$$

This contains 11.6×10^{17} particles so that one mole contains

$$\frac{11.6 \times 10^{17}}{1.9 \times 10^{-6}} = 6.0 \times 10^{23}\ \text{atoms}.$$

Question 2

(a) The vapour pressure of a solution consisting of 10 g of an involatile substance A in 75 g of propanone was 2.974×10^4 N m^{-2} at 298 K. The vapour pressure of pure propanone at this temperature is 3.055×10^4 N m^{-2}.
The vapour pressure of the solvent is given by

$$p = xp^*,$$

where x is the mole fraction of the solvent and p* is the vapour pressure of the pure solvent.
Calculate the relative molecular mass of the substance A given the following relative molecular masses: C, 12; O, 16 and H, 1.

The mole fraction, x(i), of a species, i, in a mixture is given by the equation $x(i) = \frac{n_i}{\sum_i n_i}$, where n_i is the number of moles of component i, and the summation is over all the components.

(b) By means of a microsyringe, 2×10^{-8} m^3 of a solution containing 1% by mass of the same substance A in a volatile solvent was spread on a flat water surface. After the solvent had evaporated, the area of water covered was measured and found to be 0.1 m^2.
Given that the densities of both A and the solvent are 8×10^2 kg m^{-3}, calculate the thickness of the spread layer of A.

(c) By making two further assumptions,
(i) that the molecule A is a sphere and,
(ii) that the layer is one molecule thick,
calculate the number of molecules of A on the surface. Hence calculate the number of molecules per mole (the Avogadro constant).

(d) The true value of the Avogadro constant is 6.0×10^{23} mol^{-1}. How could either of the assumptions (i) or (ii) above be modified to account for the discrepancy between the value obtained in part (c) and the correct one?

Answer

(a) The mole fraction of propanone in the liquid phase must be

$$x = \frac{p}{p^*} = \frac{2.974 \times 10^4}{3.055 \times 10^4} = 0.973.$$

The relative molecular mass of propanone, CH_3COCH_3 is

$$3 \times 12 + 16 + 6 \times 1 = 58 \text{ g}.$$

If the unknown relative molecular mass is M, then

$$x = \frac{75/58}{75/58 + 10/M}$$

so that M = 279.

(b) If both A and the solvent have the same density and the solution contains 1% of the solvent then a volume

$$\frac{2 \times 10^{-8}}{100} = 2 \times 10^{-10} \text{ m}^3$$

of A is left on the surface of the water after the solvent has evaporated. If the area covered is 0.1 m^2 the thickness of the spread layer is

$$\frac{2 \times 10^{-10}}{0.1} \text{m} = 2 \times 10^{-9} \text{ m}$$

(c) If the layer is unimolecular then the radius of the A molecules is

$$\frac{2 \times 10^{-9}}{2} = 10^{-9} \text{ m}$$

Accordingly, the volume of one molecule is

$$\frac{4}{3}\pi(10^{-9})^3$$

so that the layer contains

$$\frac{2 \times 10^{-10}}{4/3\pi(10^{-9})^3} = 4.77 \times 10^{16} \text{ molecules.}$$

The mass of this layer is

$$\begin{aligned} &= \text{density} \times \text{volume} \\ &= 8 \times 10^2 \times 2 \times 10^{-10} \\ &= 1.6 \times 10^{-7} \text{ kg.} \end{aligned}$$

It folows that in one mole of A (0.279 kg) there would be

$$8.3 \times 10^{22} \text{ molecules.}$$

(d) The true value is greater than that deduced in part (c) implying that the molecules are elongated in shape so that the number per unit area is greater and/or that the layer is larger than monolayer in thickness.

Question 3

Low Energy Electron Diffraction (LEED) can be used to study the structure of solid surfaces. Effective diffraction of the incident beam of electrons is obtained if the wavelength (λ) of the electrons is similar to the interatomic spacing. Usually the incident electrons are accelerated by an electrical potential difference of about 150 V.

The mass of an electron is 9.109×10^{-31} kg and its charge is 1.602×10^{-19} C.

In the quantum mechanical description of matter, particles (such as electrons), may be considered in terms of waves. This is wave-particle duality, and has been shown experimentally by electron (and, more recently, neutron) diffraction. The equation relating particle wavelength (λ) to the linear momentum (p) of the particle is the de Broglie relationship, $\lambda = \frac{h}{p}$.

(a) Calculate:
   (i) the kinetic energy of the incident electrons,
   (ii) the velocity of the incident electrons, and
   (iii) the wavelength of the incident electrons, given a value of 6.6×10^{-34} J s^{-1} for the Planck constant.

(b) Electrons scattered from different surface atoms in a LEED experiment, at an angle θ, may constructively interfere according to the equation

$$n\lambda = d \sin \theta$$

where λ is the wavelength of the incident electrons, d is the interatomic spacing, and n is an integer.
In a LEED experiment on MgO, the lowest diffraction angle, θ, was found to be 19.5° for electrons accelerated by a potential difference of 150 V. Calculate a value for the interionic spacing, d, on the surface of a crystal of magnesium oxide.

Answer

(a) (i) If an electron (charge 1.6×10^{-19} coulombs) is accelerated through a potential difference of 150 V then its energy is

$$150 \times 1.6 \times 10^{-19}$$
$$= 2.4 \times 10^{-17} \text{ joules.}$$

(ii) If the mass of the electron is 9.109×10^{-31} kg and v is its velocity, then

$$\text{kinetic energy} = \frac{1}{2}mv^2$$

so that

$$v = \sqrt{2 \times 2.4 \times 10^{-17}/9.109 \times 10^{-31}}$$
$$= 7.3 \times 10^6 \text{ m s}^{-1}.$$

(iii) The wavelength, λ, is given by

$$mv = \frac{h}{\lambda}$$

so that,

$$\lambda = 99.8 \text{ pm}$$

which is comparable with typical chemical bond lengths.

(b) Substituting values into the equation

$$d = \frac{n\lambda}{\sin\theta} = 2.99\ \text{Å}$$

taking $n = 1$ corresponding to the lowest diffraction angle.

Question 4

The photoelectric effect is the phenomenon whereby light, incident on a material causes electrons to be emitted, only when the frequency of the incident light is *above* a certain threshold frequency. The inability of classical mechanics to explain this phenomenon (and others), caused the formulation of quantum mechanics. The interested reader is referred to P. A. Cox, "Introduction to Quantum Theory and Atomic Structure", (OCP 37).

The photoelectric effect can be demonstrated by focusing a beam of electromagnetic radiation of variable wavelength, λ, onto the surface of caesium, and measuring the kinetic energy (KE) of the emitted electrons as a function of the wavelength. In such an experiment, the following results were obtained.

λ/nm	300	350	400	450	500	550
$10^{19} \times$ KE /J	3.20	2.27	1.53	1.01	0.56	0.12

(a) Write an expression relating the kinetic energy of the photoelectrons to the frequency of the electromagnetic radiation.

(b) Graphically determine the ionisation energy of caesium, and estimate a value of the Planck constant.

(c) What would be the velocity of the electrons if the experiment were repeated using radiation with wavelengths of 250 and 600 nm, corresponding respectively to ultraviolet light and the orange part of the visible spectrum?

Answer

(a) The relationship between the kinetic energy and the frequency of the radiation is

$$\text{KE} + \text{IE} = hc/\lambda$$

where h is the Planck constant, c is the speed of light and IE is the ionisation energy.

(b) The data shows that as the wavelength of the light is increased, the photoelectrons leave with decreasing kinetic energy. In the limit of zero kinetic energy, the energy of the incident light is just sufficient to cause ionisation. The graph in Fig. 1.2 shows this corresponds to 5.7×10^{-7} m, so that

$$\text{IE} = hc/5.7 \times 10^{-7} = 3.5 \times 10^{-19}\ \text{J}.$$

(c) For light of wavelength 250 nm the kinetic energy of the electrons would be

$$\tfrac{1}{2}mv^2 = \frac{hc}{250 \times 10^{-9}} - 3.5 \times 10^{-19} = 4.4 \times 10^{-19}\ \text{J}.$$

where v is the velocity of the emitted electrons and m is the electron mass. It follows that

$$v = \sqrt{2 \times 4.4 \times 10^{-19}/9.109 \times 10^{-31}} = 9.9 \times 10^{5}\ \text{m s}^{-1}.$$

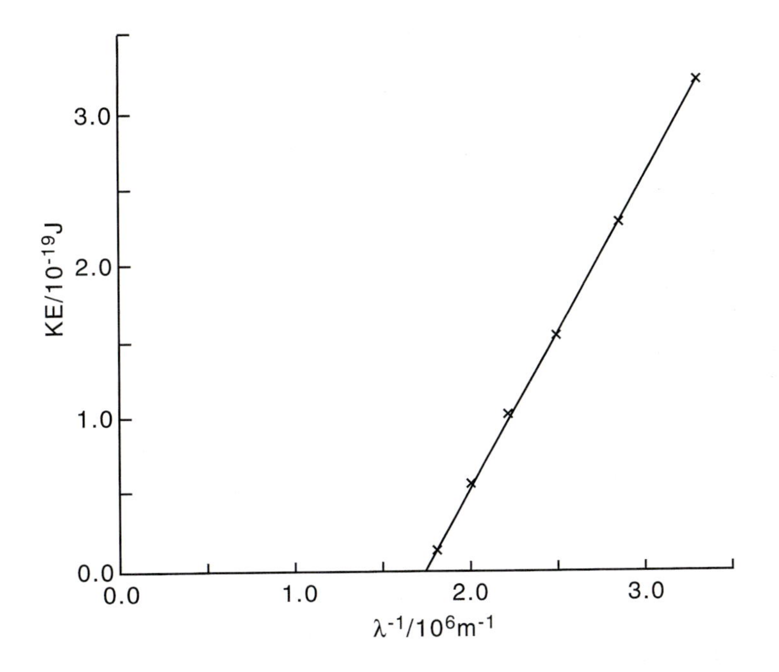

Fig. 1.2 Graph showing the determination of the caesium ionisation energy using the photoelectric effect.

With light of 600 nm, there would be insufficient energy to cause ionisation.

Question 5

The Balmer series of lines in the hydrogen atom spectrum comprises those that occur in the visible part of the electromagnetic spectrum, and have been modelled by the equation

$$\frac{1}{\lambda} = R_H\left(\frac{1}{2^2} - \frac{1}{n^2}\right)$$

where n is an integer that can take up the values 3, 4, 5, 6,...., and R_H is the Rydberg constant (= 109468 cm^{-1}).

$1\ cm^{-1} = 2.0 \times 10^{-23}$ J.

(a) Use the above formula to sketch the spectrum. Indicate on the sketch the wavelength (in nm) and the energy (in J) of the first five lines.

(b) What happens when $n = \infty$?

(c) It has been found that the wavelengths of the emission lines in the spectra of one electron atoms (H, He^+, Li^{2+} etc) are described by the equation

$$\frac{1}{\lambda} = Z^2 R_H\left(\frac{1}{n_1^2} - \frac{1}{n_2^2}\right) \quad (1.1)$$

where Z is the nuclear charge of the atom concerned, R_H is the Rydgberg constant and n_1 and n_2 are integers. If the ionisation energy of a gaseous neutral H atom is 1312.0 kJ mol^{-1}, what would be the second ionisation energy of helium?

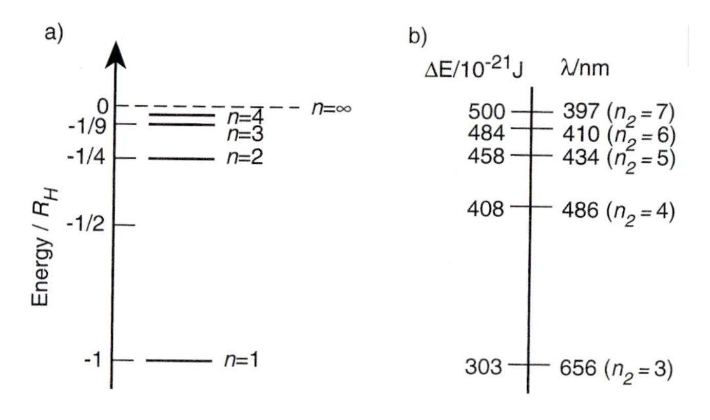

Fig. 1.3 The energy levels of the hydrogen atom and the Balmer series spectrum.

(d) Estimate the ionisation energies of the following hydrogenic ions.

$$Li^{2+}, Be^{3+}.$$

Answer

(a) Fig. 1.3 shows the energy levels of the hydrogen atom and the Balmer series spectrum.

(b) When $n = \infty$ the uppermost level in the electronic transition corresponds to the hydrogen atom in its ionised state.

(c) For ionisation of hydrogen from its ground state, $Z = 1$, $n_1 = 1$ and $n_2 = \infty$, so that the ionisation energy

$$(IE)_H - R_H = 1312 \text{ kJ mol}^{-1}.$$

For the second ionisation of helium, $Z = 2$, $n_1 = 1$ and $n_2 = \infty$. So

$$(IE)_{He} = 4R_H$$
$$= 5248 \text{ kJ mol}^{-1}.$$

(d) The ionisation energies of Li^+ and Be^{2+} must go, on the basis of equation (1.1), as the square of the nuclear charge, being respectively 9 and 16 times greater than the value for hydrogen.

Question 6

(a) What are the four quantum numbers required to specify the state of an electron in a hydrogen atom?

(b) The following wavefunctions describe the hydrogen 1s- and 2s-atomic orbitals:

$$\Psi_{1s} = Ne^{-r/a_0}, \text{ and} \qquad \Psi_{2s} = N\left\{1 - \frac{1}{2}\left(\frac{r}{a_0}\right)\right\}e^{-r/2a_0}$$

where $a_0 = \dfrac{\varepsilon_0 h^2}{\pi m_e e_2}$ and is the so-called Bohr radius (the radius of the hydrogen atom as given by Bohr's electrostatic model of the atom).

(i) Sketch the form of the two wavefunctions, and find the location of the node in the 2s-orbital in terms of a_0.

(ii) Explain what is meant by the radial distribution function (rdf) of an electron, and sketch the form of the rdf for the 1s- and 2s-orbitals given above.

(iii) What is the most probable distance (expressed in terms of a_0) of a 1s-electron from the nucleus in a hydrogen atom?

Answer

(a) The four quantum numbers are summarised in Table 1.1

Table 1.1 Quantum Numbers

Name	Possible Values	Information
principal, n	1, 2, 3, ...	energy and size
orbital, l	0, 1, 2, ..., n − 1	shape
angular momentum, m_l	−l, −l +1, ..., l − 1, l	orientation
spin, m_s	$-\frac{1}{2}, \frac{1}{2}$	spin

(b) (i) The two wavefunctions are sketched in Fig 1.4. The node in the 2s orbital occurs where $\Psi_{2s} = 0$ which is when r = 2a.

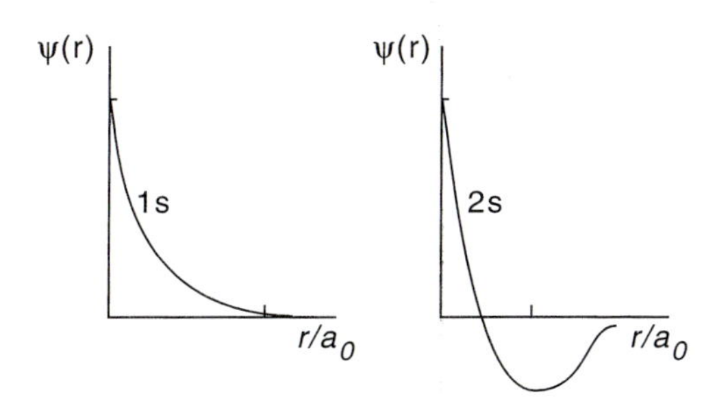

Fig. 1.4 The 1s and 2s wavefunctions of the hydrogen atom.

(ii) The square of the wavefunction, Ψ^2 at any point is a measure of the probability of finding the electron at that point. It follows that for spherically symmetrical wavefunctions,

$$\int_0^\infty 4\pi r^2 \Psi^2 dr = 1$$

since the probability of finding an electron somewhere must be unity. The quantity

$$r^2\Psi^2$$

is known as the radial distribution function and gives the probability of finding the electron at *any* distance r from the nucleus. These radial distibution functions are shown in Fig.1.5.

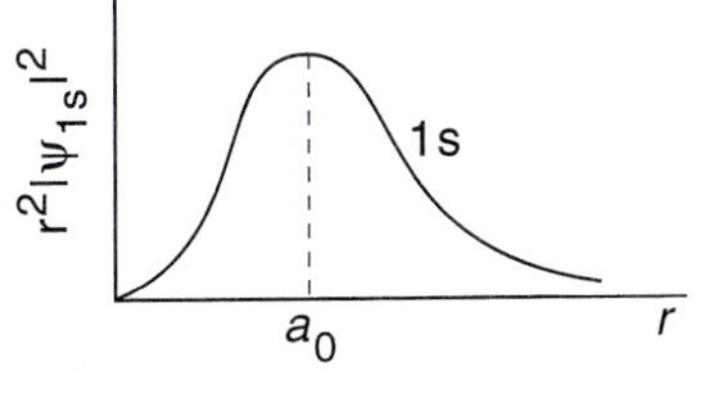

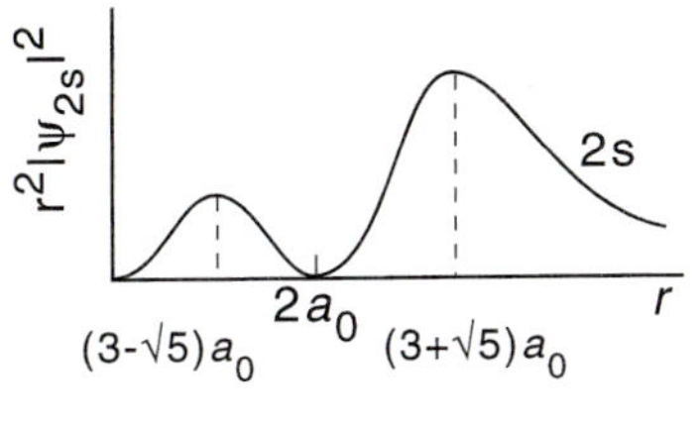

Fig. 1.5 The 1s and 2s radial distribution fuctions for the hydrogen atom.

(iii) Maxima (or minima) in the radial distribution function must occur when

$$\frac{d[r^2\Psi^2]}{dr} = 0$$

For the 1s orbital,

$$\frac{d}{dr}[r^2N^2e^{-2r/a_0}]$$
$$= N^2\left[2re^{-2r/a_0} - \frac{2r^2}{a_0}e^{-2r/a_0}\right].$$

This is zero when

$$2r - \frac{2r^2}{a_0} = 0$$

so that $r = 0$ or a_0. The former is a minimum, the latter a maximum.

Question 7

The ionisation energies of hydrogen (13.6 eV) and of oxygen (13.5 eV) are approximately the same. Why is this?

Answer

The ionisation energies of the elements hydrogen to neon are plotted in Fig. 1.6. There is a general increase across a period. The simple linear trend for the second row is disrupted at Li, B and O implying either that He, Be and N are particularly stable and/or Li, B and O are particularly unstable. These observations may be understood as follows.

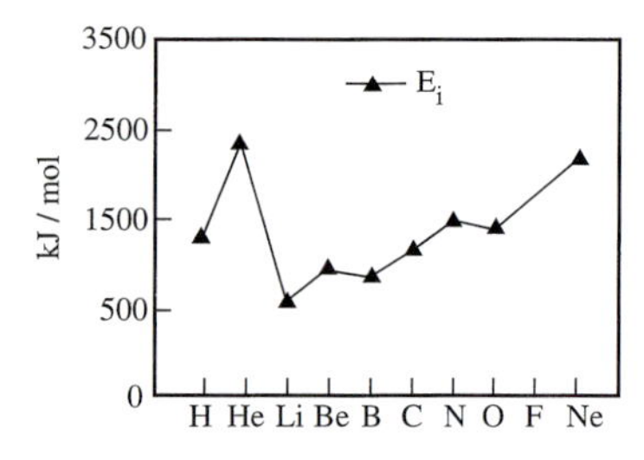

Fig. 1.6 The trend in ionisation energies of the elements H to Ne.

- The change in ionisation energy between He and Li is due to the fact that the electron to be ionised from Li is in a 2s orbital whereas that from He is in a 1s orbital. Being so much closer to the nucleus, the helium electron is held much more tightly.
- The Be/B step correlates with the point where the last electron to be added was 2s and the next is 2p and we know that 2p electrons are attracted less strongly to the nucleus than are 2s electrons, which penetrate closer in and also shield the 2p electrons from the nucleus.
- At the N/O step, N has a half-filled sub-shell and O has two electrons spin-paired in a 2p orbital and two half-filled 2p orbitals. The double occupancy of an orbital causes some degree of repulsion between electrons which is unfavourable; in addition, half-filled (and filled) sub-shells are also special in that they have a maximum amount of energy of what is known as *exchange energy*, which is a quantum mechanical contribution to the stability from, electrons with the same spin in degenerate orbitals.

Question 8

Explain the trend in electron affinity (EA) observed for the halogens given below.

Halogen:	F	Cl	Br	I
EA / kJ mol^{-1}	328	349	325	296

Answer

The electron affinity of an atom is the energy released when an electron is added to the atom in the gaseous phase.

The decrease in electron affinity from Cl to Br to I is expected since the p orbitals accepting the added electron become larger and the electron is less tightly bound to the nucleus. The value for fluorine is however lower than that seen for chlorine: because of the exceptionally small size of the fluorine atom, the entering electron causes an anomalously large increase in the repulsive energy among all the valence-shell electrons.

The atomic radii of the halogens are as follows: F, 0.57 Å; Cl, 0.98 Å; Br, 1.12 Å; I, 1.32 Å.

Question 9

Give the ground electron configurations of the following chemical species.

$$O,\ H^-,\ O^{2-},\ Mg^{2+},\ Cu^{2+},\ Au,\ Ga^+,\ Mn,\ Br^+,\ Te^{2-}.$$

Answer

The ground electron configurations are:

O	(He) $2s^2 2p^4$
H^-	$1s^2$
O^{2-}	(He) $2s^2 2p^6$
Mg^{2+}	(Ne)
Cu^{2+}	(Ar) $3d^9$
Au	(Xe) $4f^{14} 5d^{10} 6s^1$
Ga^+	(Ar) $4s^2 3d^{10}$
Mn	(Ar) $4s^2 3d^5$
Br^+	(Ar) $3d^{10} 4s^2 4p^4$
Te^{2-}	(Kr) $4d^{10} 5s^2 5p^6$,

where the bracketted terms denote the noble gas cores for the different species.

Question 10

The effective nuclear charge (Z_{eff}) observed in atoms heavier than hydrogen (many-electron atoms) can be modelled by the equation

$$Z_{eff} = Z - \sigma$$

where Z is the true nuclear charge and σ is an empirical parameter known as the shielding (or screening) constant. Analysis by Slater of a large number of compounds caused him to formulate a set of rules which calculate, to a good first approximation, the value of the shielding parameter for any electron in any atom. As applied to an outermost electron, Slater's rules are as follows.

(1) Divide the electron configuration of the atom (or ionic species) into sets as follows.

[1s]; [2s, 2p]; [3s, 3p]; [3d]; [4s, 4p]; [4d]; [4f];...........

(2) Unfilled orbitals do not contribute to the shielding parameter.

(3) Electrons in the set containing the outermost electron contribute 0.35 to the shielding parameter, except if the set is 1s, whereby the shielding parameter takes the value 0.30. The outermost electron itself contributes zero.

(4) Electrons in the set immediately to the left of the set containing the outermost electron contribute 0.85 to σ if they are in s or p orbitals, and 1.00 if they are in d or f orbitals.
(5) Electrons in sets that are two or more to the left of that containing the outermost electron each contribute 1.00 to the value of σ.

(a) Explain why the "shielding effect" occurs in many-electron atoms.
(b) Estimate the effective nuclear charge for the following.
 (i) A 2p electron of carbon.
 (ii) A 2p electron of nitrogen.
 (iii) A 2p electron of oxygen.
 (iv) A 2p electron of fluorine.
(c) Comment on the periodic trend in values obtained in part (b).

Answer

(a) Electrons in an atom can "shield" (or screen) each other from the full positive charge of the atomic nucleus. This causes a weakening in the electrostatic force of attraction between the shielded electron and the atomic nucleus. This weakening can be interpreted as an effective nuclear charge (Z_{eff}) felt by the electron. It is evident that electrons in orbitals which approach closer to the nucleus are more effective at shielding than are less "penetrating" orbitals. For example, the screening of 2p electrons by 2s electrons is greater than by other 2p electrons.
(b) (i) The ground state electron configuration of carbon is $1s^22s^22p^2$. Applying Slater's rules gives

$$\sigma = (3 \times 0.35) + (2 \times 0.85) = 2.75,$$

which implies

$$Z_{eff} = 3.25.$$

(ii) Nitrogen has the electron configuration $1s^22s^22p^3$. Consequently, the shielding parameter for the outermost 2p electron is 3.1, giving

$$Z_{eff} = 3.9.$$

(iii) Oxygen, $1s^22s^22p^4$ has a shielding parameter equal to 3.45 with

$$Z_{eff} = 4.55.$$

(iv) Fluorine, $1s^22s^22p^5$ with its shielding parameter of 3.8 gives rise to

$$Z_{eff} = 5.20.$$

(c) The effective nuclear charge increases across the first short period reflecting the imperfect screening characteristic of the 2p orbitals. Unsurprisingly, the electronegativity and electron affinity increase from C to N to O to F.

2 Molecules: the beginning of chemistry

2.1 Aims

This chapter seeks to provide questions which cover both core knowledge in, and some more advanced concepts relating to, the following topics.

- The different types of bonding including ionic, covalent, metallic, van der Waals and hydrogen bonds.
- The prediction of molecular shape using the Valence Shell Electron Pair Repulsion (VSEPR) rules.
- The use of molecular orbital theory in explaining covalent bonding.
- The use of different types of spectroscopy (infrared, nuclear magnetic resonance (NMR), ultraviolet/visible, photoelectron) applied to the understanding of bonding.
- The application of spectroscopy, and of mass spectrometry in determining the structure of molecules.

2.2 Worked examples

Question 1

(a) Briefly summarise the different types of bonding found in the following solid state species.

(i) MgO; (ii) SiO_2; (iii) H_2O; (iv) Cl_2; (v) Fe.

(b) The densities of diamond and ice are 3.5 and 0.94 g cm^{-3} respectively. Both solids adopt analogous structures. In diamond the C-C bond length is 0.155 nm, whereas in water the O-O distances are close to 0.276 nm. Comment.

Answer

(a) (i) MgO: ionic bonding. The rocksalt structure is adopted (see Fig. 3.5).

(ii) SiO_2: covalent bonding. The giant lattice is formed from corner sharing SiO_4 tetrahedra.

(iii) The H_2O molecules are covalently bonded within themselves but hydrogen bonded to each other (see Fig. 2.1(a)).

(iv) The covalently bonded Cl_2 molecules are held within the solid lattice by van der Waals forces.

(v) Iron usually adopts the body-centred cubic structure. (Fig 1.1(a)). At higher temperatures iron takes up the face-centred cubic structure (Fig 1.1(b)). As the temperature increases even more this rearranges to again give the body-centred cubic structure.

The Atomium structure in Brussels (Fig 2.2) is based on a body-centred cubic lattice. The "atoms" contain science exhibitions whilst the "bonds" are escalators.

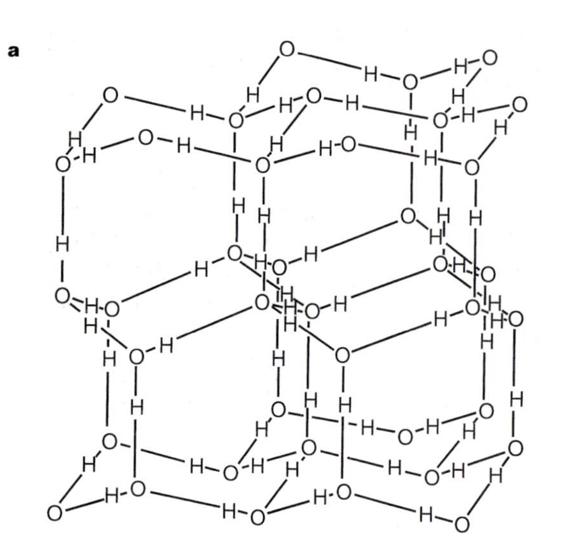

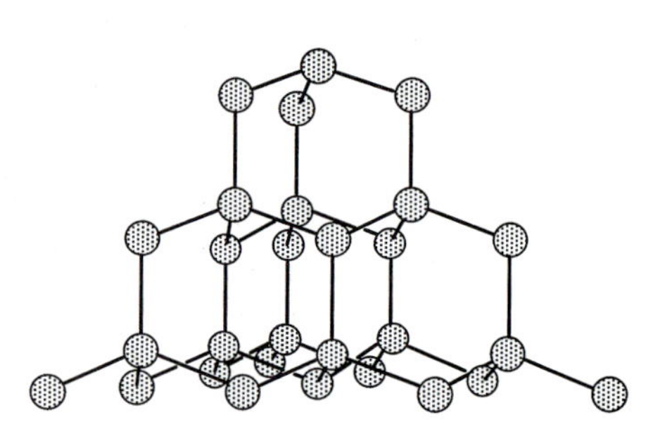

Fig. 2.1 Figure showing the structures of (a) ice and (b) diamond.

(b) The structures of the two solids are shown in Fig. 2.1. There is four-co-ordination around each C or O and the similarity of the lattices is evident. Density is defined as mass per unit volume. The relative molecular mass of water is 18, whereas that of carbon is 12. It would be expected that the densities, ρ, are in the ratio

$$\rho(\mathrm{C}) : \rho(\mathrm{H_2O}) = \frac{12}{(0.155)^3} : \frac{18}{(0.276)^3}$$
$$= 3.76 : 1$$
$$= 3.5 : 0.94.$$

It can be seen that the lattice spacing as determined by X-ray crystallography is consistent with that implicit in the relative densities of the two materials. The much lower density of ice as compared to diamond can be seen to derive from the substantially greater O-O distance. This reflects the fact that the ice lattice is held together by hydrogen bonds which are significantly weaker and longer than the covalent C-C bonds in diamond.

Question 2

Predict and explain the shape of the following molecules or ion in the gaseous state.

(a) BeH_2; (b) BF_3; (c) NH_4^+; (d) PCl_5; (e) SF_6; (f) IF_7.

Notice the question asks about the molecular geometries in the *gas* phase. Be aware that changes can occur in the solid state. For example solid PCl_5 exists as $[PCl_4]^+ [PCl_6]^-$!

Answer

The structures can be predicted using the VSEPR method. This comprises a set of empirical (experiment based) rules, which are as follows.

Fig. 2.2 The Atomium; (a) from afar and (b) close up.

- Atoms in a molecule are bound together by electron pairs. Each pair comprises a molecular bond. In some cases more than one bond may bind any two atoms together, so creating double or triple bonds.
- Some atoms in a molecule may also possess pairs of electrons which are not involved in bonding. These are so called lone pairs (or non-bonded pairs).
- The bonding pairs and lone pairs around any particular atom in a molecule adopt a distribution (shape) that minimises the repulsion between the pairs of electrons, both lone and bonding.
- If there is more than one kind of electron pair, the hierarchy of repulsion is:

For more details on VSEPR theory see M. J. Winter, "Chemical Bonding", (OCP 15).

$$\text{lone pairs} > \text{triple bonds} > \text{double bonds} > \text{single bonds.}$$

In general the calculation of the number of bonding and lone pairs requires knowledge of the number of valence electrons in the constituent atoms. This is readily discovered from a knowledge of the location of the atoms in the Periodic Table of elements. All chemists need to be instantly familiar with at least the s and p block elements. If the reader is not, appropriate action is recommended!

Lone pairs are thought to be most repulsive as they are contracted towards the nucleus and so occupy a greater solid angle than bonding pairs. This means that they keep as far apart as possible.

In the questions posed, all the molecules are of the general form AZ_n where A is a central atom bound to n (= 2, 3, 4, 5, 6 or 7) identical "ligands", Z. The structures adopted, both as observed experimentally and in accordance with the VSEPR rules, are the regular geometries shown in Fig. 2.3. These shapes are described respectively as linear (BeH_2), trigonal (BF_3), tetrahedral (NH_4^+), trigonal bipyramidal (PCl_5), octahedral (SF_6) and pentagonal bipyramidal (IF_7).

Question 3

Predict and contrast the shapes of the following three structures containing nitrogen:

$$NH_4^+;\ NH_3;\ NH_2^-.$$

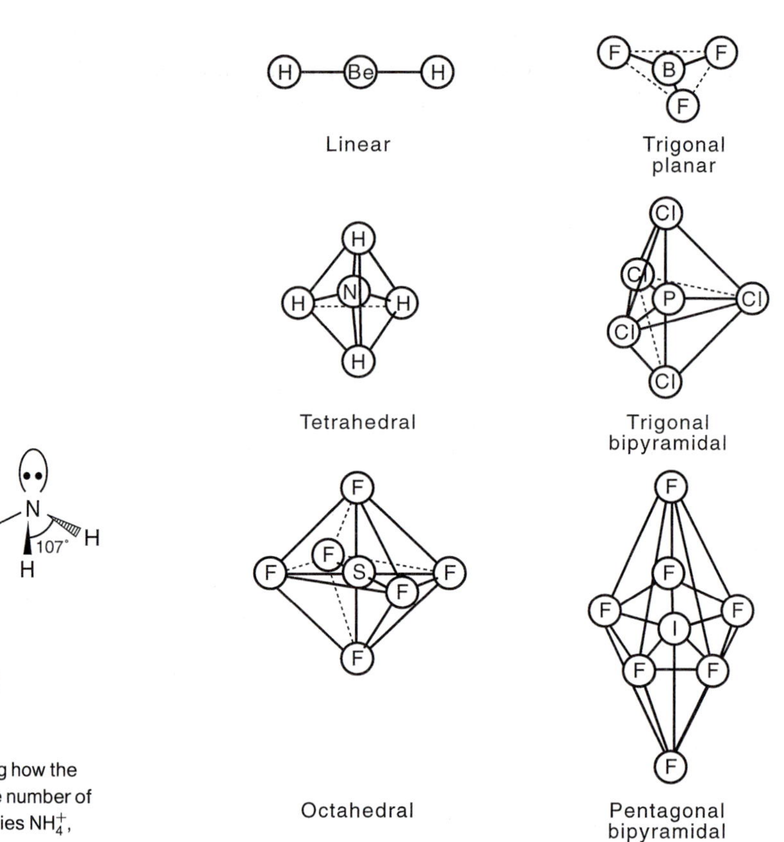

Fig. 2.3 The linear (BeH_2), trigonal planar (BF_3), tetrahedral (NH_4^+), trigonal bipyramidal (PCl_5), octahedral (SF_6) and pentagonal bipyramidal (IF_7) structures.

Fig. 2.4 A diagram showing how the bond angle decreases as the number of lone pairs increase in the series NH_4^+, NH_3 and NH_2^-.

The presence of three bonds and one lone pair of electrons in NH_3 but two bonds and two lone pairs in NH_2^- can be appreciated in the "dot and cross" Lewis structures.

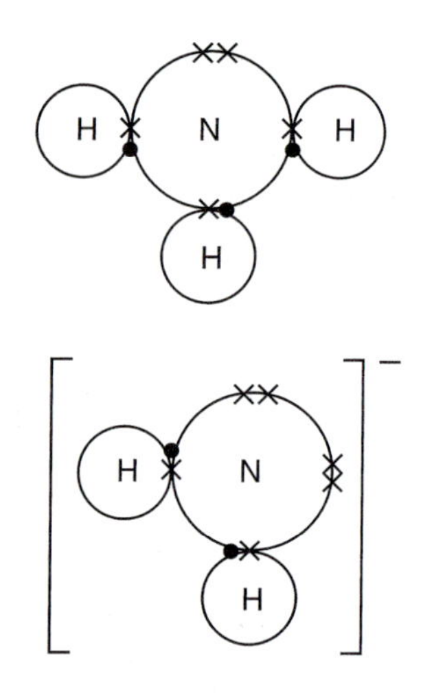

Answer

The case of the ammonium ion, NH_4^+, was discussed in Question 2; there are four identical N-H bonds distributed tetrahedrally around the central nitrogen atom (Fig. 2.4). The H-N-H bond angles are close to 109.5° (the "tetrahedral angle"). For ammonia, NH_3, there are three bonding pairs of electrons and one lone pair. So to a first approximation the four bonds are tetrahedrally disposed around the nitrogen. However as the lone pair is more strongly repelling than the N-H bonds, there is a (small) distortion from the ideal shape, so that the H-N-H bond angle is reduced to 107°.

The distorted tetrahedron is the structure adopted by many molecules with the AZ_3L formula where L denotes a lone pair. Other examples, together with the Z-A-Z bond angle include the following: NF_3 (102°); PH_3 (94°); PF_3 (98°); PCl_3 (100°); PBr_3 (102°) and PI_3 (102°).

In the amide ion, NH_2^-, there are two bonding pairs and two lone pairs of electrons surrounding the central nitrogen atom. The structure is again approximately tetrahedral but distorted by the stronger repelling pair of the, now two, lone pairs so that the H-N-H bond angle is smaller than in NH_3.

The shape shown in Fig. 2.4 for the NH_2^- ion is characteristic of many AZ_2L_2 systems as illustrated in Fig. 2.5.

The bond angle in the NH_2^- ion varies with environment. The value of 104° in Fig. 2.4 refers to solid potassium amide KNH_2. In $NaAl(NH_2)_4$ it is as low as 94°.

Question 4

Predict and contrast the shapes of the following three molecules.

$$PF_5;\ SF_4;\ ClF_3.$$

Answer

Phosphorous pentachloride was discussed in Question 2: there are five bonds surrounding the central phosphorous atom, but no lone pairs. The trigonal bipyramidal structure of the molecule, PF_5, is shown in Fig. 2.6, from which it can be seen that the P-F bonds are almost all exactly the same length. The two very slightly longer bonds are described as "axial", whereas the three P-F bonds around the midriff of the molecule are labelled as "equatorial".

If the reader is unsure about the number of lone pairs, it is advisable to construct a dot and cross Lewis structure.

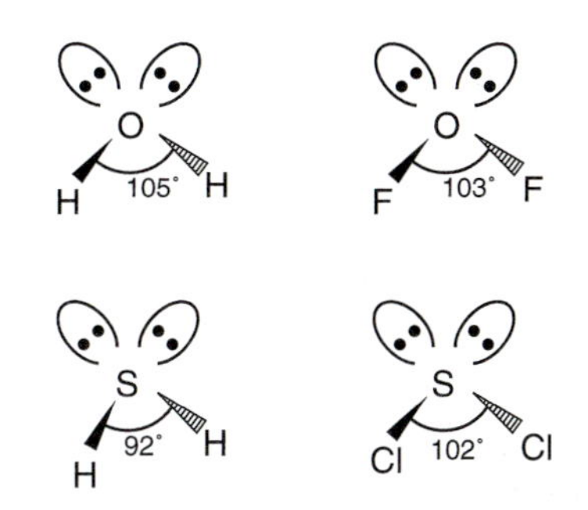

Fig. 2.5 The structures and bond angles of OH_2, OF_2, SH_2, and SCl_2.

The molecule SF_4 has the central sulphur atom surrounded by four bonding pairs (S-F bonds) and one lone pair of electrons. To a first approximation therefore the structure is trigonal bipyramidal. However at this point, a dilemma is encountered when applying the VSEPR rules: should the lone pair be placed in an axial or in an equatorial position? The two possibilities are shown in Fig. 2.7. To decide the structure of lowest energy, we must invoke the "repulsion hierarchy":

lone pairs > single bonds.

Retaining for present purposes the idealised trigonal bipyramidal structure and tabulating the different possible interactions in terms of the angles between the pairs of electrons, denoting bonding pairs as BP and lone pairs as LP, we find:

Structure	90°	120°	180°
a	3 × BP-LP	3 × BP-BP	1 × BP-LP
	3 × BP-BP		
b	2 × BP-LP	1 × BP-BP	1 × BP-BP
	4 × BP-BP	2 × BP-LP	

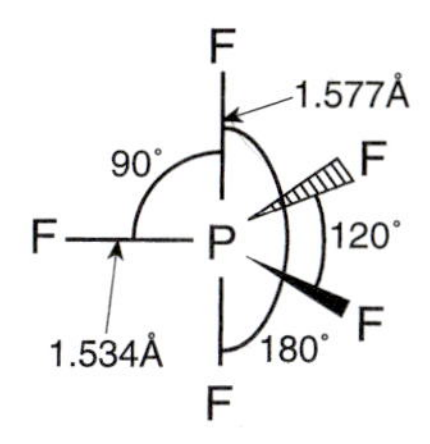

Fig. 2.6 The structure and bond angles of PF_5.

If the repulsion between pairs is greatest when the angle between them is least, we should expect the 90° interactions to play a dominant role in determining the molecular geometry. This *is* the case, so that structure (b) in Fig. 2.7 is more stable than structure (a) since it only has two interactions (as opposed to three) of the lone pair-bonding pair type; the lone pair of electrons in SF_4 adopts an equatorial position.

To summarise the above, in AZ_4L type molecules such as SF_4, the lone pair adopts an equatorial rather than axial position in the idealised trigonal bipyramidal structure on account of the greater repelling power of the lone pair as compared to bonding pairs. Having decided on the location of the lone pair in the idealised structure of SF_4, the last step in prediction of the true structure is to consider the distortion of the ideal caused by the repulsion hierarchy. The

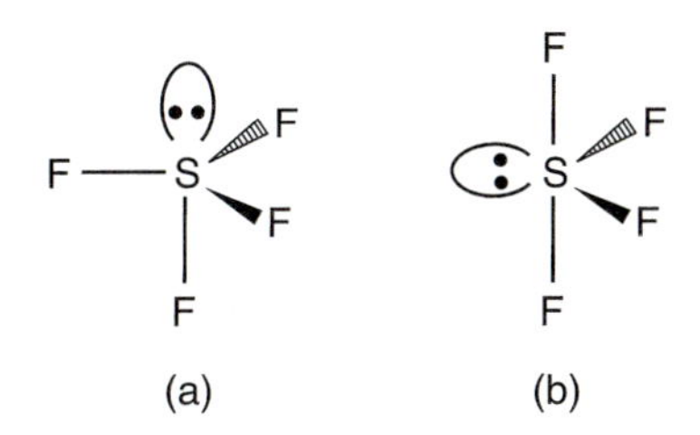

Fig. 2.7 The two possible structures for SF_4. The lone pair can be either (a) axial or (b) equatorial.

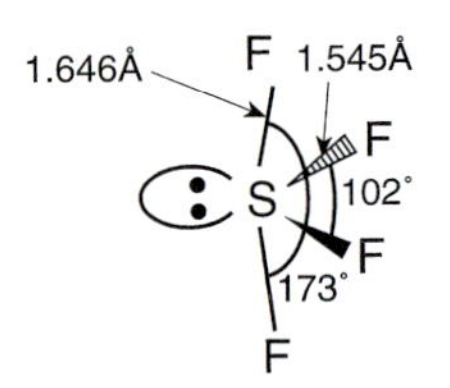

Fig. 2.8 The observed shape of SF_4.

experimentally found shape of SF_4 is shown in Fig. 2.8, where the stronger repulsion of the lone pair over the bonding pairs is again evident.

Turning to the case of ClF_3, it can be appreciated that the central chlorine atom is surrounded by three bonding pairs and two lone pairs. The idealised shape is therefore again trigonal bipyramidal. As with SF_4, however, it is impossible to immediately decide whether the lone pairs adopt axial or equatorial positions within the idealised structures. The three possibilities are illustrated in Fig. 2.9. To decide on the lowest energy of the three possible

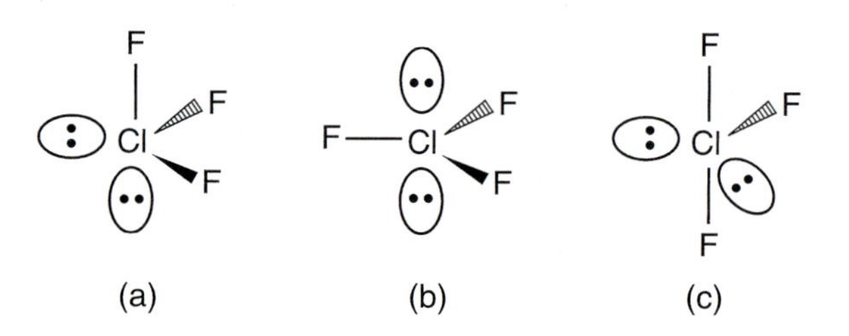

Fig. 2.9 The three possible stuctures of ClF_3: (a) lone pairs in both equatorial and axial positions, (b) two lone pairs in the axial positions or (c) both lone pairs in equatorial positions.

bonding pair/ lone pair distributions for ClF_3 within the trigonal bipyramidal structure, we again tabulate the different possible types of interactions as follows.

Structure	90°	120°	180°
a	1 × LP-LP	1 × BP-BP	1 × BP-LP
	3 × BP-LP	2 × BP-LP	
	2 × BP-BP		
b	6 × LP-BP	3 × BP-BP	1 × LP-LP
c	4 × BP-LP	1 × LP-LP	1 × BP-BP
	2 × BP-BP	2 × BP-LP	

Applying the repulsion hierarchy:

$$\text{LP-LP} > \text{LP-BP} > \text{BP-BP},$$

we can appreciate that structure (a) is unfavourable because of a LP-LP interaction at an angle of just 90°. This is therefore the least stable structure. Comparing structure (b) and (c) we see that the former has six LP-BP interactions whereas the latter has just four LP-BP repulsions but two, more favourable, BP-BP repulsions at angles of 90°. The interactions at this angle dominate since this corresponds to the closest separation of the mutually repelling electron pairs. So structure (c) is the least energetic and most stable; both lone pairs in ClF_3 occupy equatorial positions.

Fig. 2.10 Experimentally determined bond angles and structure of ClF_3.

Another molecule of the AZ_3L_2 form which takes up a distorted trigonal bipyramid structure with two equatorial lone pairs is BrF_3. The F-Br-F angle is 172°.

Last, we need account for the distances in the idealised trigonal bipyramidal distribution of bonds and the Cl atoms in ClF_3 as induced by the repulsion hierarchy. Fig. 2.10 shows the structure of the molecule as experimentally determined: it can be seen that the lone pairs repel the bonding pairs so that the axial to axial F-Cl-F angle is compressed from 180° to 175°.

Question 5

Predict and contrast the shapes of the following molecules:

$$SF_6; BrF_5; XeF_4.$$

Answer

As introduced in Question 2, SF_6 has six coordinating pairs of electrons surrounding the central sulphur atom which adopt an octahedral geometry.

The anion PF_6^- has a similar structure to SF_6.

The molecule BrF_5 is of the type AZ_5L, having five bonding pairs and one lone pair of electrons. The idealised shape of the molecule is that shown in Fig. 2.11. Note that as the distribution of electrons pairs has octahedral symmetry, all six locations around the Br atom, are, in the idealised structure, equivalent: there is no equatorial versus axial distinction as encountered in the trigonal bipyramidal geometry of AZ_5 molecules.

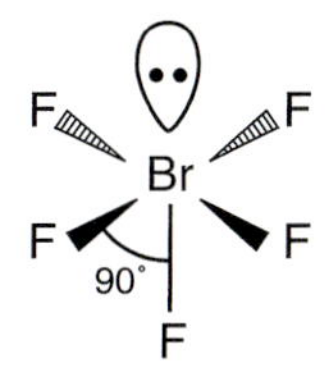

Fig. 2.11 The idealised shape of BrF_5.

Experimental results on BrF_5 show that the distortion expected on the basis of the repulsion hierarchy,

$$\text{lone pair} > \text{bonding pairs},$$

occurs as shown in Fig. 2.12, and the F-Br-F bond angle shown is 85° rather than the 90° expected for the idealised structure.

ClF_5 adopts a structure similar to BrF_5.

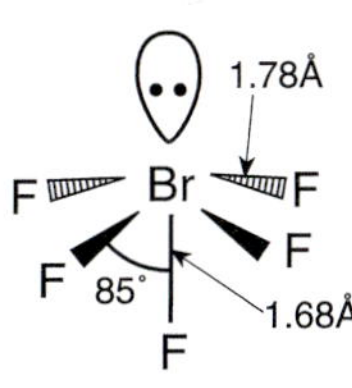

Fig. 2.12 The experimentally determined structure of BrF_5.

Turning to the case of XeF_4, this is of the type, AZ_4L_2 with four bonding pairs and two lone pairs of electrons. The idealised structure is based on an octahedral distribution of electron pairs. Fig. 2.13 shows the two possible arrangements of the lone pairs described as "cis" or "trans". It is clear that the lone pair-lone pair

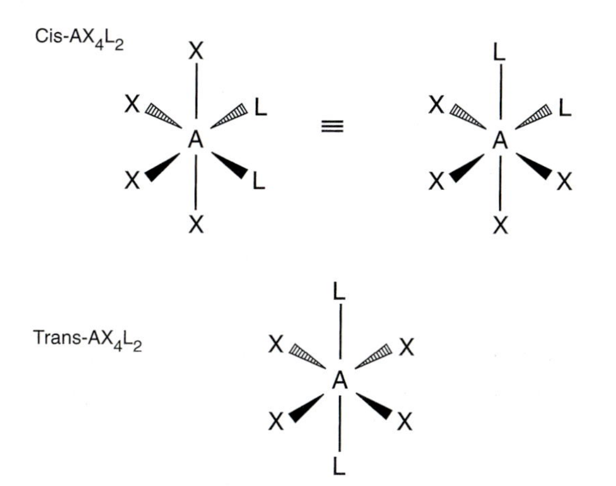

Fig. 2.13 The (a) cis and (b) trans structures of AZ_4L_2 (L = lone pair).

repulsion, which is greater than lone pair-bonding pair or bonding pair-bonding pair repulsions, will be least in the trans arrangement so that XeF_4 is planar with the two lone pairs above and below the square plane (Fig. 2.14).

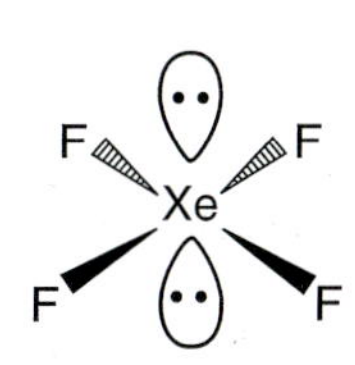

Fig. 2.14 The structure of XeF_4.

Molecules adopting a similar shape as XeF_4 are the anions ICl_4^- and BrF_4^- both of which are planar with two trans lone pairs.

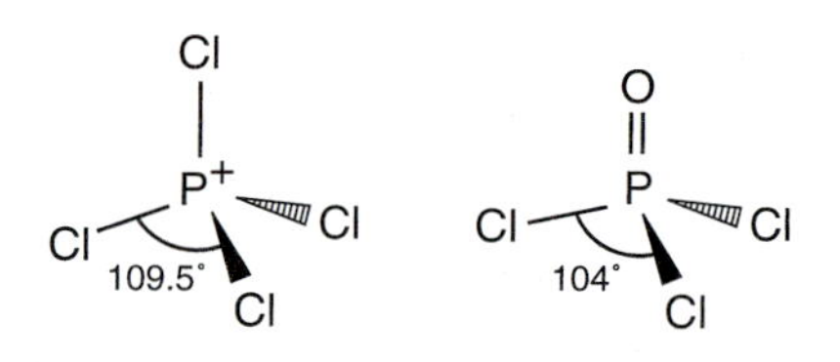

Fig. 2.15 A double bond compresses bond angles.

Question 6

Predict the shapes of the following molecules.

$$PCl_4^+;\ OPCl_3.$$

Answer

The PCl_4^+ cation is surrounded by four bonding pairs of electrons but no lone pairs. The shape of the molecule is therefore tetrahedral (see Fig. 2.15).

$OPCl_3$ is surrounded by five bonding pairs of electrons, two of which form a double bond to the oxygen atom. The idealised distribution of the four ligands is therefore tetrahedral, but invoking the repulsion hierarchy,

double bonds > single bonds,

it can be appreciated that the Cl-P-Cl bond angle is compressed below the ideal tetrahedral angle as seen, for example, in PCl_4^+ to 104°.

OPF_3 takes up a similar structure to $OPCl_3$ (Fig. 2.15) with a closely similar F-P-F angle near 104°.

Question 7

Predict the shapes of the following molecules.

$$SF_4;\ SOF_2;\ SOF_4;\ S_2F_{10};\ S_2F_2.$$

Answer

In SF_4, the central sulphur atom is surrounded by four bonding pairs of electrons and one lone pair. The structure was described and discussed in Question 4.

In SOF_2, sulphur is surrounded by four bonding electron pairs, of which two form a double bond to oxygen, and two bond singly to fluorine, leaving a lone pair of electrons. The structure is therefore based on a distorted tetrahedron as shown in Fig. 2.16, in which the repulsion hierarchy,

lone pairs > double bonds > single bonds

can be used to account for the F-S-F bond angle of 93° being significantly less than the tetrahedral angle (strong repulsion of the S-F bonding electrons pairs by both the lone pair and the double bond) and the F-S-O bond angle of 107° which is below, but not so far removed from, the tetrahedral angle (repulsions by a lone pair being greater than by a double bond).

In SOF_4, there are six bonding pairs of electrons of which two form a double bond to oxygen. The five ligands adopt an approximately trigonal bipyramidal distribution with the S=O double bond occupying an equatorial position for analogous reasons to the lone pair in SF_4 being located in a similar position (Fig. 2.16). This again reflects the repulsion hierarchy

double bonds > single bonds.

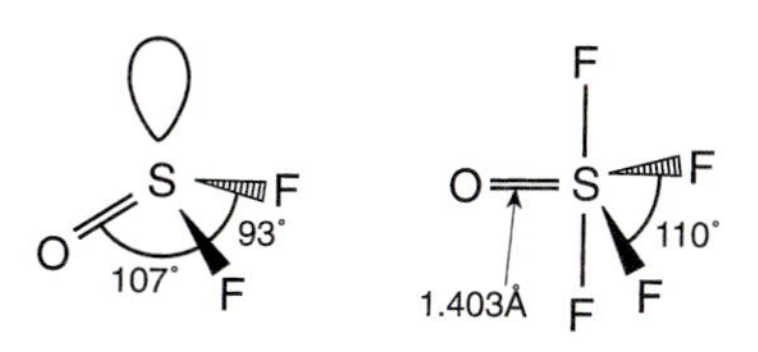

Fig. 2.16 The structures of SOF_2 and SOF_4 as determined by electron diffraction.

The structure of S_2F_{10} is based on two octahedra sharing one vertex (apex) as shown in Fig. 2.17.

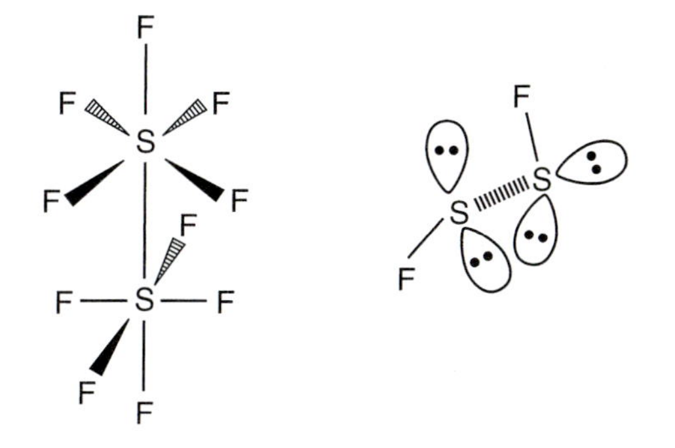

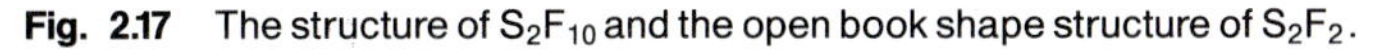
Fig. 2.17 The structure of S_2F_{10} and the open book shape structure of S_2F_2.

In S_2F_2, each sulphur is surrounded by two bonding pairs of electrons (forming S-S and S-F bonds respectively) and two lone pairs. The approximate distribution of electron pairs around each is therefore tetrahedral. Note that the molecule is *not* planar but twisted (in an "open book" form) to minimise the repulsions between the lone pairs. S_2Cl_2, F_2O_2 and H_2O_2 (Fig. 2.18) adopt analogous structures.

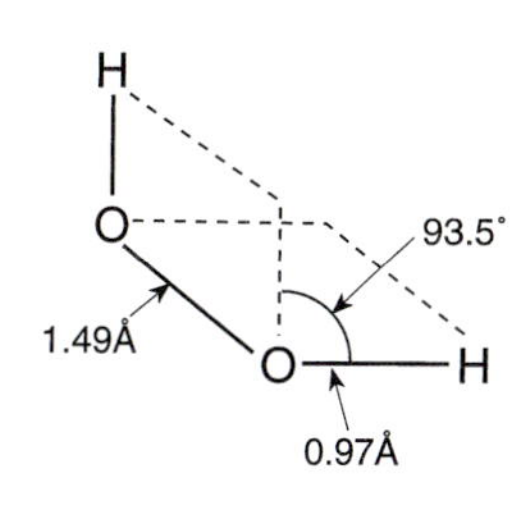

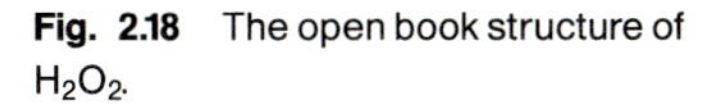
Fig. 2.18 The open book structure of H_2O_2.

Question 8

The bond enthalpies of H_2^+, H_2 and He_2^+ are 255, 432 and 322 kJ mol^{-1} respectively. Comment.

Answer

The data can be understood by constructing a molecular orbital energy level diagram. The constituent atomic orbitals are the two 1s orbitals on each of the component atoms. Fig. 2.19 shows the orbitals and the associated energy level diagram.

In H_2^+, there is just one electron which occupies the lowest orbital which is bonding in character as it has a lower energy than the hydrogenic 1s orbitals. The bond order of H_2^+ is $\frac{1}{2}$. H_2 has two electrons in the bonding orbital and a bond order of one. For He_2^+, of the three electrons, two will again occupy the bonding orbital but the third must be accommodated in the antibonding orbital. This is a direct consequence of the aufbau, or building-up, principle which comprises three rules.

Bond order (the number of bonds) is defined to be $\frac{1}{2}$\{(no. of bonding electrons) – (no. of antibonding electrons)\}.

(1) Only two electrons of opposite spin can be accommodated in an orbital (the so called Pauli Exclusion principle).

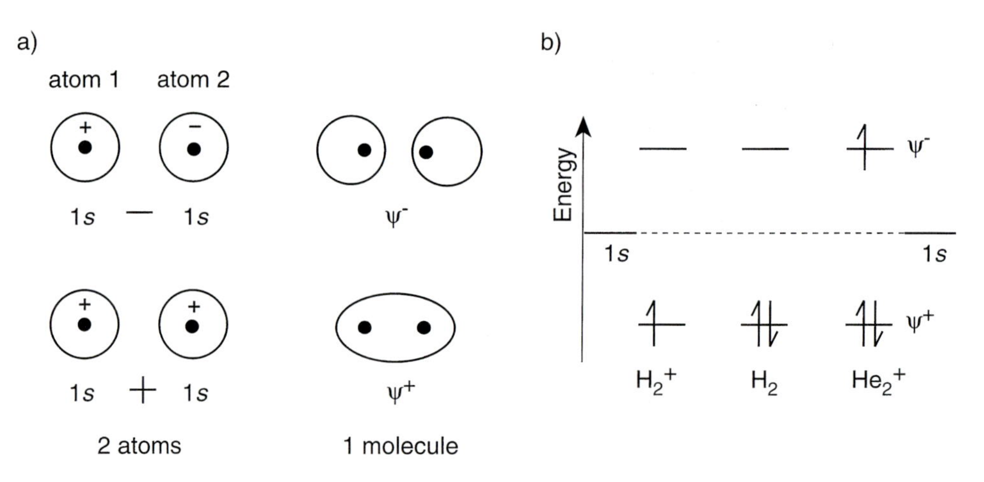

Fig. 2.19 (a) First row diatomic molecular orbital diagram schematically illustrating the shapes of the orbitals. (b) First row diatomic energy level diagram with electron occupancy of the ground state indicated by arrows. Note that the energy scale is arbitrary and not the same for the three species (the 1s orbital of He is lower than that for H and the number of electrons affects the bond strength).

(2) Electrons are assigned to orbitals one at a time with the next electron being assigned to the lowest energy orbital available (not already occupied by two electrons).

(3) If the lowest energy orbitals available are degenerate (have the same energy) then one electron is first assigned to each orbital with the spin of all the single electrons being the same. Then a second electron of opposite spin is assigned to each orbital in turn.

The first two rules suffice to show He_2^+ must have a bond order of $1-\frac{1}{2}=\frac{1}{2}$. The third rule will be encountered in subsequent questions.

Question 9

(a) Consider a homonuclear diatomic molecule in which the constituent atoms have 2s and 2p valence orbitals.
- (i) Describe the molecular orbitals which can arise from interaction of the 2s orbitals on the two atoms.
- (ii) Describe the molecular orbitals which can arise from interaction of the 2p orbitals on the two atoms.
- (iii) Draw a labelled molecular orbital scheme for a homonuclear diatomic molecule, neglecting interaction between 2s and 2p orbitals. Use this scheme to predict the number of unpaired electrons in O_2.
- (iv) What would be the effect on the interatomic distance in the O_2 molecule of reduction to O_2^- or oxidation to O_2^+?

(b)
- (i) How would the molecular orbital scheme in (a)(iii) be modified if interactions between 2s and 2p orbitals were taken into account?
- (ii) Draw labelled molecular orbital scheme for B_2, noting that the molecule contains two unpaired electrons.
- (iii) What would be the effect on the interatomic distance in B_2 of reduction to B_2^- or oxidation to B_2^+?

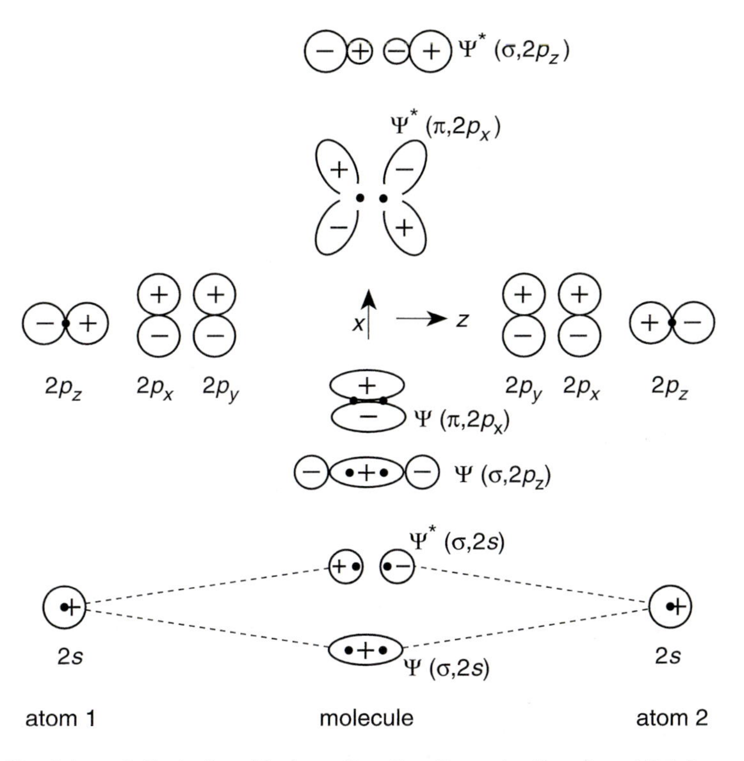

Fig. 2.20 Schematic illustration of the formation of bonding and antibonding orbitals for second row diatomic molecules. The two atoms are shown schematically on either side of the *molecule* that they form. There are $\Psi(\pi, 2p_y)$ and $\Psi^*(\pi, 2p_y)$ orbitals degenerate with the corresponding $\Psi(\pi, 2p_x)$ and $\Psi^*(\pi, 2p_x)$ orbitals but orientated pointing into and out of the page.

Answer

(a) (i) The formation of a bonding, Ψ, and an antibonding, Ψ^*, orbital from the interaction of two 2s atomic orbitals is shown in Fig. 2.20.

(ii) Fig. 2.20 shows the overlap of the 2p orbital leads to σ and π bonds. Of the former type, there is one bonding, $\Psi(\sigma, 2p_z)$, and one antibonding, $\Psi^*(\sigma, 2p_z)$ orbital. Of the π-type, there are two bonding, $\Psi(\pi, 2p_x)$ and $\Psi(\pi, 2p_y)$, and two antibonding, $\Psi^*(\pi, 2p_x)$ and $\Psi^*(\pi, 2p_y)$, orbitals.

(iii) The molecular orbital scheme for a homonuclear diatomic molecule, neglecting the interaction between 2s and 2p orbitals is shown in Fig. 2.21(a). The number of electrons corresponding to O_2 are shown. It can be seen that in the latter case each bonding π orbital is filled with a pair of electrons but that the two π^* orbitals each contain a single electron in accordance with the rules associated with the aufbau principle given in the previous question. O_2 therefore has two unpaired electrons and as a result displays the phenomenon of paramagnetism—attraction into a magnetic field. Most materials show diamagnetism—they are repelled by magnetic

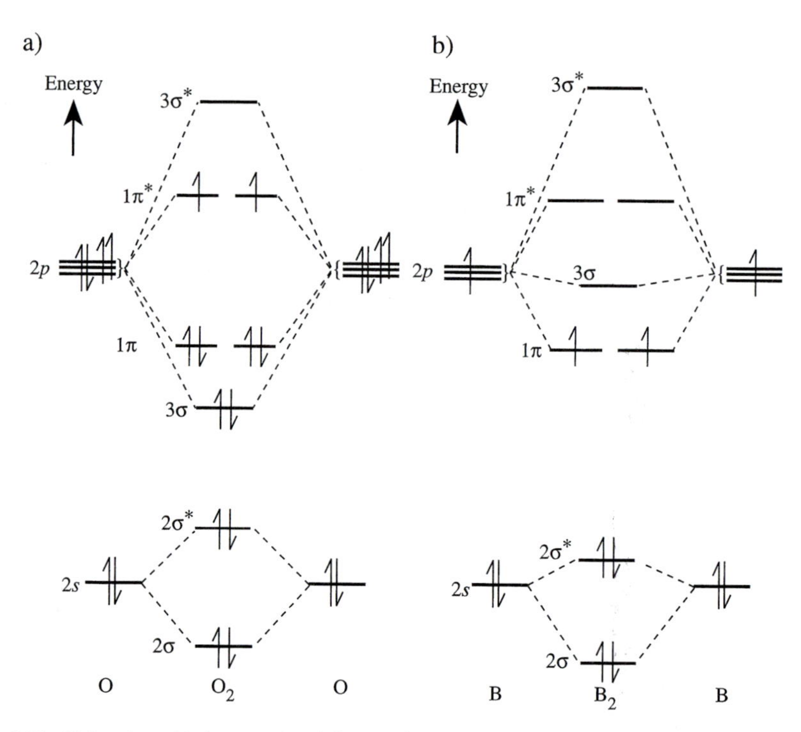

Fig. 2.21 Molecular orbital energy level diagram for a second row diatomic molecule (a) in the absence of s / p mixing (showing the electron assignment for O_2) and (b) with s / p mixing (showing the electron assignment for B_2). σ and π orbitals are numbered in order; 1 σ and 1 σ^* are not shown.

The two unpaired electrons in O_2 molecule cause it to exhibit radical type properties. An example will be encountered in Question 15 of chapter four in the oxidation of hydrocarbons.

fields. This latter behaviour is associated with fully filled orbitals when all electrons exist in spin-opposed pairs.

(iv) The introduction of a further electron into O_2 to form O_2^- would add a third electron into the π^* orbitals. These are antibonding, so the bond length would be expected to increase. Conversely, on removing one electron from O_2 to form O_2^+, the loss of an antibonding electron (π^*) would strengthen the bond, causing the bond length to decrease.

The observed bond lengths (nm) are:

The O - O bond length data relates to the solids O_2PtF_6, O_2, KO_2 and Na_2O_2.

O_2^+	O_2	O_2^-	O_2^{2-}
0.113	0.121	0.133	0.149

(b) (i) The molecular orbital energy scheme shown in Fig. 2.21(a) applies to O_2 and F_2 of the homonuclear diatomics of the first short period. However, for Li_2-N_2, the scheme shown in Fig. 2.21(b) applies. This arises since for O_2 and F_2, the 2s-2p energy gap is large, as the higher nuclear charge results in the shielding of the 2p electrons by the 2s electrons being very effective. In contrast, for Li-N, this gap is less so that there is mixing of the s and the $2p_z$ orbitals in forming the σ bonds. The result of this is that the 2σ and $2\sigma^*$ molecular orbitals are lower in energy, whilst the 3σ and $3\sigma^*$ are higher. Comparison of (a) and (b) in Fig. 2.21 illustrates this point.

(ii) The electron arrangement for B_2 is given in Fig. 2.21(b). Notice that since the 1π orbitals are of lower energy than the 3σ orbitals, the two most energetic electrons occupy separate π orbitals and the B_2 molecule has two unpaired electrons.

(iii) The addition of an electron to B_2 to form B_2^- would add an electron to a bonding 1π orbital so that the bond length would be expected to shorten. In contrast, removal of a bonding 1π electron from B_2 to create B_2^+ would lengthen the B-B bond.

Question 10

Draw a molecular orbital energy level diagram for CO. How does it differ from that for N_2? The bond dissociation enthalpy of N_2 is 942 kJ mol^{-1} and that for CO is 1072 kJ mol^{-1}. Explain this observation.

Answer

The energy level diagram for CO is shown in Fig. 2.22. The pattern of σ and π molecular orbitals resembles that in N_2 but since the constituent atomic

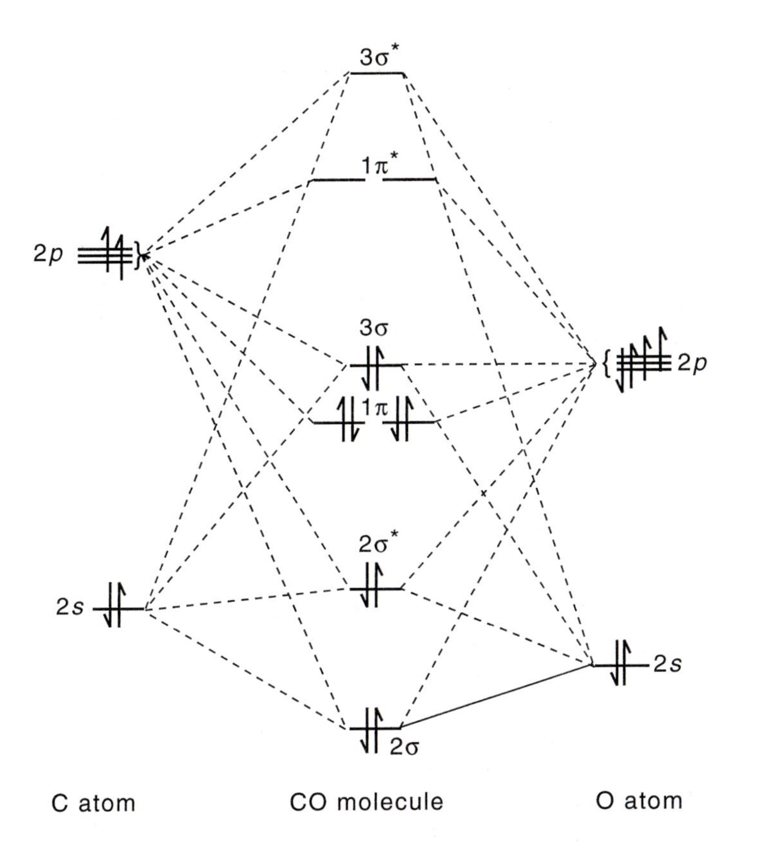

Fig. 2.22 The molecular orbital diagram for carbon monoxide.

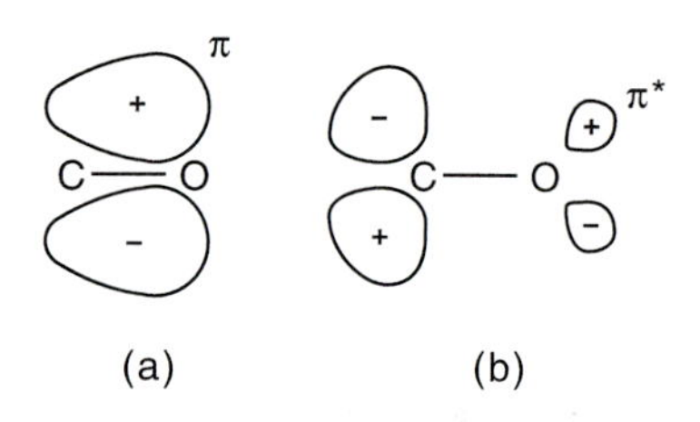

Fig. 2.23 Diagrammatic sketches of the π and π^* molecular orbitals in carbon monoxide: (a) 1π bonding orbital and (b) $1\pi^*$ antibonding orbital.

orbitals (2s, 2p) on oxygen are lower in energy than those on carbon, the bonding molecular orbitals resemble (in shape and energy) the atomic orbitals on oxygen more than they do those of carbon. Conversely, the antibonding orbitals resemble more the carbon atomic orbitals (Fig. 2.23).

Question 11

The photoelectron spectrum of oxygen was determined experimentally using electromagnetic radiation with wavelength, $\lambda = 58.4$ nm. Peaks in the resulting spectrum were found, amongst others, at electron kinetic energies near 9.3 and 5.2 eV. Explain the origins of these peaks. To what ionisation energies do these peaks correspond?

Answer

The photons used to record the photoelectron spectra have a wavelength of 58.4 nm corresponding to an energy,

$$E = \frac{hc}{\lambda} = 3.4 \times 10^{-18}\ \mathrm{J}$$

where h is the Planck constant and c is the speed of light. Since energy is conserved,

$$E = IE + KE$$

where IE is the ionisation energy and KE is the kinetic energy. The latter has values of 9.3 and 5.2 eV corresponding respectively to 1.5×10^{-18} and 0.8×10^{-18} J. Accordingly the corresponding IE values are 1.9×10^{-18} and 2.6×10^{-18} J (c.a 11.9 and 16.3 eV). The origin of the peaks can be understood with reference to the molecular orbital diagram of O_2 in Fig. 2.21(a) where the outermost electrons are in the $1\pi^*$ and 1π orbitals. The peaks correspond to ionisation from these levels.

Question 12

The photoelectron spectrum of HCl is shown in Fig. 2.24. Comment.

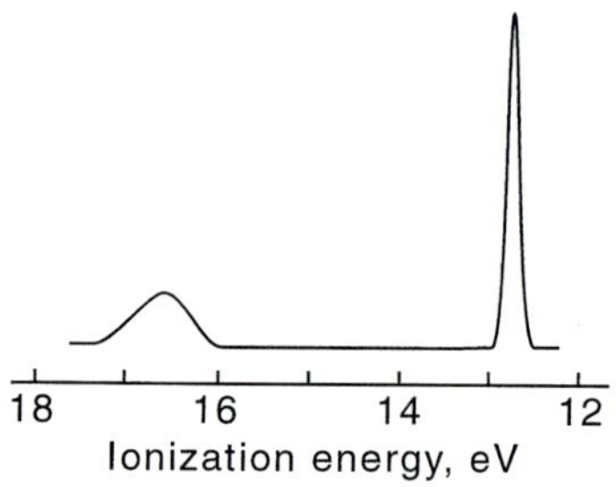

18 16 14 12
Ionization energy, eV

Fig. 2.24 A low resolution photoelectron spectrum of HCl.

Answer

In HCl a simple molecular orbital description of the bonding suggests that the H 1s orbital and the Cl 3p orbital pointing at the H atom will mix effectively to form a (σ) bonding and an (σ^*) antibonding molecular orbital. However, the other two Cl 3p orbitals are non-bonding (Fig. 2.25).

The lowest energy ionisation in Fig. 2.25 therefore corresponds to loss of a photoelectron from the non-bonding orbitals. The ionisation at higher energy is that from the bonding σ orbital.

Question 13

Aqueous solutions of phenolphthalein at different pH values have been studied by taking their ultraviolet/visible absorption spectra, Fig. 2.26.

(a) Why is phenolphthalein useful as a weak acid-strong base titration indicator?

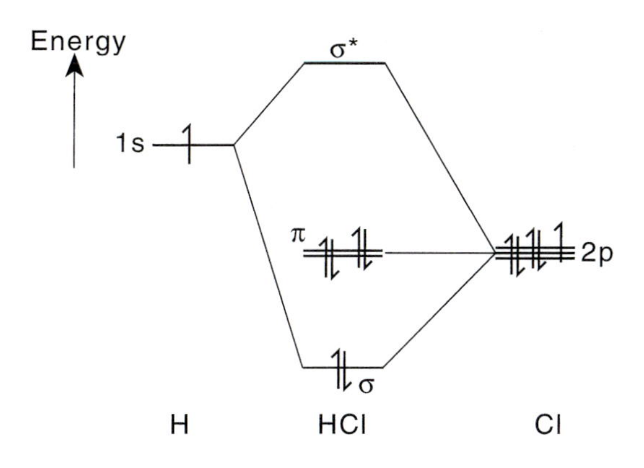

Fig. 2.25 The molecular orbital diagram for HCl.

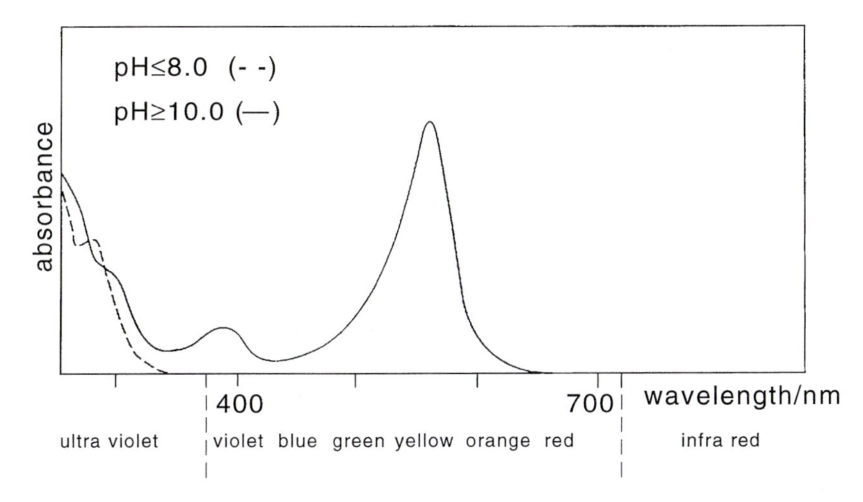

Fig. 2.26 An ultraviolet / visible spectrum of phenolphthalein in aqueous solution at pH ≤ 8 and pH ≥ 10.

(b) What happens to the electrons in phenolphthalein when the molecule absorbs ultraviolet/visible radiation.

(c) What would be the effect on the absorption spectrum of changing the concentration of phenolphthalein?

Answer

(a) The spectrum shows that at pH 8 and below phenolphthalein only absorbs light in the ultraviolet region and so is colourless to the human eye. However at pH 10 and above the molecule absorbs light in the yellow part of the spectrum and so appears coloured (pink). The colour change occurs because the molecule becomes deprotonated at the higher pH.

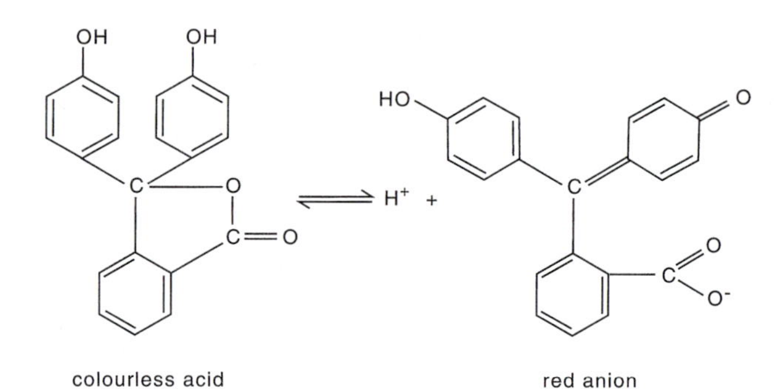

(b) The energy of the ultraviolet/visible light is used to excite an electron in the absorbing molecule to a higher energy level. In the case of phenolphthalein this is likely to cause the promotion of a delocalised π electron into an empty π^* orbital.

(c) Increasing the concentration of the phenolphthalein would cause a stronger absorbance of light. Quantitatively,

Equation (2.1) is the Beer-Lambert law discussed further in C. E. Wayne and R. P. Wayne, "Photochemistry", (OCP 39).

$$\text{Absorbance} = \varepsilon c l \tag{2.1}$$

where c is the concentration, l is the pathlength of the light through the sample and ε is the so called extinction coefficient of the molecule in question.

Question 14

Molecular vibration is stimulated by the absorption of infra-red (IR) radiation.

(a) There are four types of vibration modes in carbon dioxide. Discuss their origins.

(b) The IR spectrum of CO_2 only shows two absorption bands. Why is this?

(c) A pure liquid with molecular formula C_2H_6O is analysed using infra-red spectrometry. Part of its IR spectrum is shown in Fig. 2.27. Use Table 2.1 to answer the following.

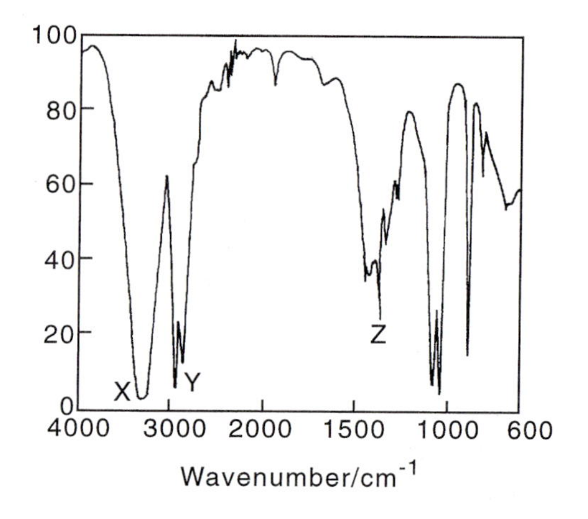

Fig. 2.27 The infra red spectrum of the compound C_2H_6O.

Table 2.1 Characteristic infrared frequencies.

Functional group	Type of vibration	Wavenumber / cm^{-1}
Alkanes	C-H stretch	2850–2960
	C-H bend	1370–1480
Alcohols, phenols	O-H stretch	3610–3640
	O-H stretch (hydrogen bonded)	3200–3600
Carboxylic acids	O-H stretch (hydrogen bonded)	2500–3000
	C=O stretch	1700–1725
Esters	C=O stretch	1735–1750
Aldehydes	C=O stretch	1720–1740
Ketones	C=O stretch	1705–1725

Data adapted from G. Aylward and T. Findlay, "SI Chemical Data", 3rd Edn., John Wiley, Brisbane (1994).

(i) Which bonds in the compound are responsible for the bands X, Y and Z.
(ii) Why is peak X broad?

Answer

(a) There are two vibrating modes of the linear CO_2 molecule: two degenerate bends, a symmetric stretch and an asymmetric stretch.

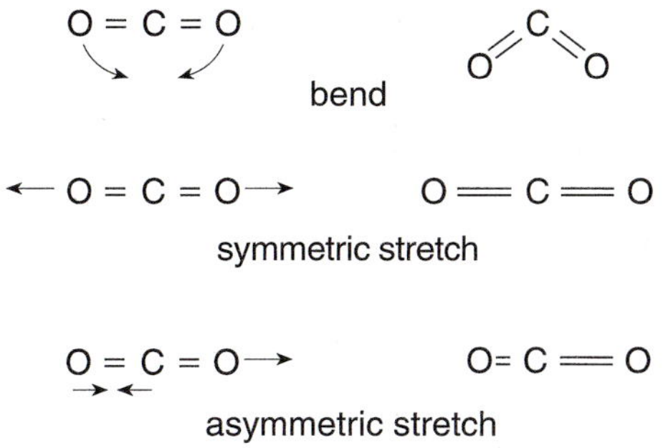

O = C = O

bend

← O = C = O →   O ═ C ═ O

symmetric stretch

O = C = O →   O= C ═ O

asymmetric stretch

(b) The two absorbance bands arise from excitation of the bending and asymmetric stretching modes in CO_2. No peak is seen for the symmetric stretch since this vibration involves no change in dipole moment so that it is impossible for the infrared radiation to excite this vibration. This is an example of a *selection rule* operating: IR active vibrations must have a change in dipole moment.

The origin of spectroscopic selection rules is discussed by J. M. Brown, "Molecular Spectroscopy", (OCP 55).

(c) (i) The peaks X, Y and Z correspond respectively to a O-H stretch, a C-H stretch and a C-H bend. These observations suggest the liquid is ethanol, C_2H_5OH.
(ii) The breadth of peak X arises since the O-H group participates in intermolecular hydrogen bonding.

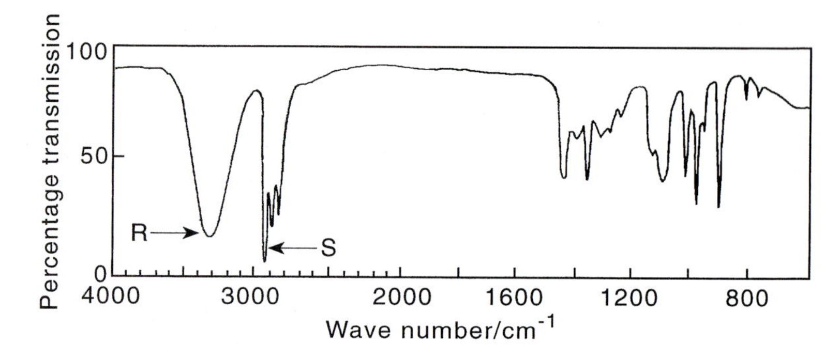

Fig. 2.28 The infrared spectrum of $C_4H_{10}O$.

Question 15

Fig. 2.28 shows the IR spectrum for a compound of molecular formula $C_4H_{10}O$.

(a) Which bonds are responsible for the absorption peaks labelled R and S?

(b) Draw all possible structural isomers of this compound.

Answer

(a) The peaks R and S can be attributed to O-H and C-H bonds.

(b) There are four possible structural isomers:

$CH_3CH_2CH_2CH_2OH$,

$CH_3CH_2CH(OH)CH_3$,

$(CH_3)_3COH$, and

$(CH_3)_2CHCH_2OH$.

Question 16

(a) Outline the principles of relative atomic mass determination by mass spectrometry.

(b) Calculate the relative atomic mass of magnesium given that the mass spectrum shows three peaks corresponding to ^{24}Mg, ^{25}Mg and ^{26}Mg in the ratio 63 : 8.1 : 9.1.

(c) Why is it not possible to distinguish between the ions $^{16}O^+$ and $^{32}S^{2+}$ in a low resolution mass spectrometer?

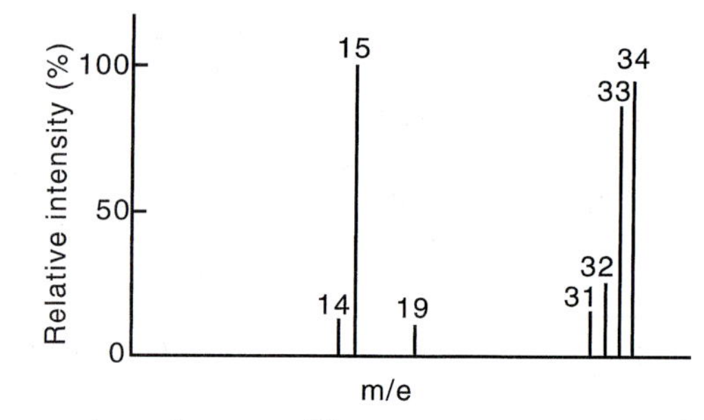

Fig. 2.29 The mass spectrum of compound X.

(d) Identify giving reasons, the compound, X, which gives the mass spectrum in Fig. 2.29 from the list below

$$CH_4, H_2O_2, NH_3, CH_3F, PH_4$$

assuming that the only isotopes of elements concerned are ^{1}H, ^{12}C, ^{14}N, ^{16}O, ^{19}F and ^{31}P and that no mass peaks occur at masses greater than 35.

Answer

(a) A schematic diagram showing a mass spectrometer is given in Fig. 2.30. In finding the relative atomic mass of an element, a sample is first vaporised and positive ions are then produced from the vapour. The latter are then accelerated by a known electric field, deflected by a magnetic field and finally detected. Only ions with a particular mass to charge ratio reach the detector for given value of the magnetic and electric fields.

(b) To find the RAM of magnesium it is necessary to take a weighted average of the isotopic masses:

$$\text{RAM} = \frac{(63 \times 24) + (8.1 \times 25) + (9.1 \times 26)}{(63 + 8.1 + 9.1)}$$

$$= 24.3.$$

(c) $^{16}O^{+}$ and $^{32}S^{2+}$ have the same charge to mass ratio and so a low resolution mass spectrometer does not distinguish between the two species.

(d) The relative molecular masses of the candidate molecules are: CH_4, 16; H_2O_2, 34; NH_3, 17; CH_3F, 34; PH_3, 34. Since a parent peak appears at a mass to charge ratio of 34 the molecules CH_4 and NH_3 can be immediately discounted.

Consideration of the other peaks suggests plausible molecule fragments if CH_3F is responsible for the spectrum: CH_3F^+, 34; CH_2F^+, 33; CHF^+, 32; CF^+, 31; F^+, 19; CH_3^+, 15 and CH_2^+, 14. In the case of PH_3, the only likely products are PH_3^+, 34; PH_2^+, 33; PH^+, 32 and P^+, 31 so this choice of molecule would be inconsistent with the data given. Likewise, H_2O_2 would probably give peaks at $H_2O_2^+$, 34; HO_2^+, 33; O_2^+, 32; OH^+, 17; O^+, 16.

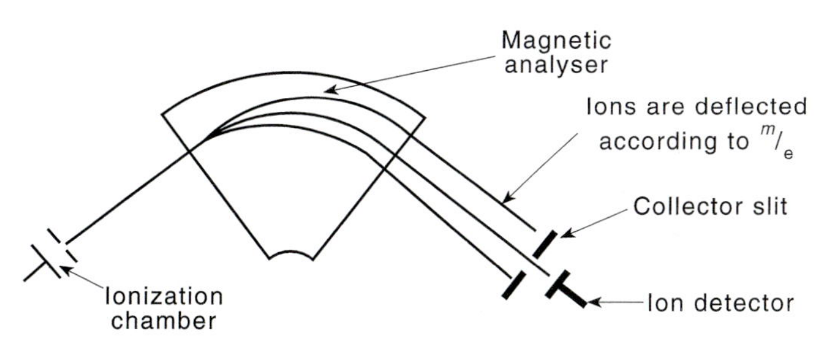

Fig. 2.30 A mass spectrometer.

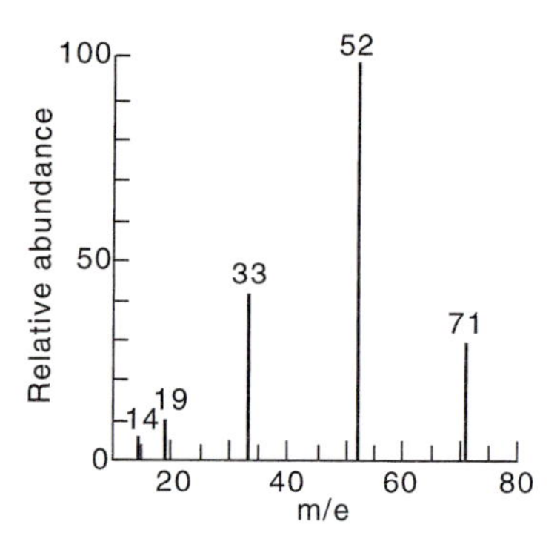

Fig. 2.31 The mass spectrum of the unknown compound in Question 17.

Question 17

The mass spectra of a compound known to contain nitrogen and atoms of only one other element is shown in Fig. 2.31. Identify the compound.

Answer

The molecule is NF_3 consistent with the observations of peaks in the mass spectrum at m/e values of NF_3^+, 71; NF_2^+, 52; NF^+, 33; F^+, 19; N^+, 14.

Question 18

(a) The mass spectrum of chlorine shows peaks at masses 70, 72, and 74. The heights of the peaks are in the ratio 9 : 6 : 1. What is the relative abundance of ^{35}Cl and ^{37}Cl?

(b) A mixture of chloroalkanes was analysed by means of a mass spectrometer. Four peaks were observed, each corresponding to a molecular ion, designated A, B, C, and D in the table below.

A	Relative Molecular Mass	50	52		
	Intensity	0.0300	0.0100		
B	Relative Molecular Mass	84	86	88	
	Intensity	0.180	0.120	0.020	
C	Relative Molecular Mass	98	100	102	
	Intensity	0.072	0.048	0.008	
D	Relative Molecular Mass	118	120	122	124
	Intensity	0.216	0.216	0.072	0.008

Deduce with reasoning;
(i) possible formulae for A, B, C and D, and
(ii) the relative amounts of the four compounds in the mixture.

(c) When the compound $CHBrCl_2$ was analysed in the mass spectrometer it gave four peaks corresponding to its molecular ions, whose masses and intensities are given in the table below.

Relative Molecular Mass	162	164	166	168
Intensity	0.09	0.15	0.07	0.01

Use the value obtained for the abundance of the chlorine isotopes to calculate the masses and relative abundances of the two bromine isotopes and explain the pattern of intensities observed.

Answer

(a) The relative abundances of ^{35}Cl and ^{37}Cl is 3 : 1. This gives a probability ratio of 9 : 6 : 1 for the molecules $^{35}Cl_2$, $^{35}Cl^{37}Cl$ and $^{37}Cl_2$ respectively.

(b) (i) Molecules A has only two relative molecular masses differing by two units so contains just one chlorine atom per molecule. The relative molecular masses suggest the molecule is CH_3Cl. The intensity ratio of 3 : 1 is consistent with part (a).

Molecule B shows three molecular masses each differing by two units so contains two chlorine atoms. The relative molecular masses suggest the molecule is CH_2Cl_2. The intensity ratio of 0.18 : 0.12 : 0.002 or 9 : 6 : 1 is again consistent with the data in part (a).

Molecule C similarly contains two chlorine atoms and the relative molecular masses implies the molecular formula is $C_2H_4Cl_2$. This molecule has two isomers: 1,1-dichloromethane and 1,2-dichloromethane. The data given in the question does not permit a distinction to be made between these two possibilities.

Molecule D has four possible molecular masses, again each differing by two mass units, suggesting it contains three chlorine atoms. The relative molecular masses imply the structure is $CHCl_3$. The intensity pattern is in the binomial ratio

$$3^3 : 3 \times 3^2 : 3 \times 3 : 1$$

or $$27 : 27 : 9 : 1.$$

(ii) The relative amounts are reflected by the sum of the intensities for each molecule so that the total ratio A : B : C : D is
(0.03 + 0.01) : (0.18 + 0.12 + 0.12) : (0.72 + 0.048 + 0.048) : (0.216 + 0.216 + 0.082 + 0.008)
or 0.04 : 0.032 : 0.128 : 0.512
or 1 : 8 : 32 : 128.

(c) Given the data in parts (a) and (b) above there must be two isotopes of bromine with masses 79 and 81 so that the four peaks correspond to

162, $CH^{79}Br^{35}Cl_2$; 164, $CH^{79}Br^{35}Cl^{37}Cl$ and $CH^{81}Br^{35}Cl_2$;
166, $CH^{79}Br^{37}Cl_2$ and $CH^{81}Br^{35}Cl^{37}Cl$; 168, $CH^{81}Br^{37}Cl_2$.

Comparing the intensities of the lightest and heaviest peaks, these are in the ratio of 9 : 1 expected for the relative amounts of ^{35}Cl : ^{37}Cl if the abundances of ^{79}Br and ^{81}Br are equal. The intensity of the other two peaks can be shown to be consistent with this inference. The four intensities are in the ratio

$$9 : (6 + 9) : (1 + 6) : 1$$

or $$0.09 : 0.15 : 0.07 : 0.01$$

Question 19

(a) The three isomers shown in Fig. 2.32 each have a molecular ion with m/e = 72 in their mass spectra. The low resolution NMR spectrum of one of these isomers is given in Fig. 2.33. The number of hydrogen atoms associated with each peak in the NMR spectrum is labelled above

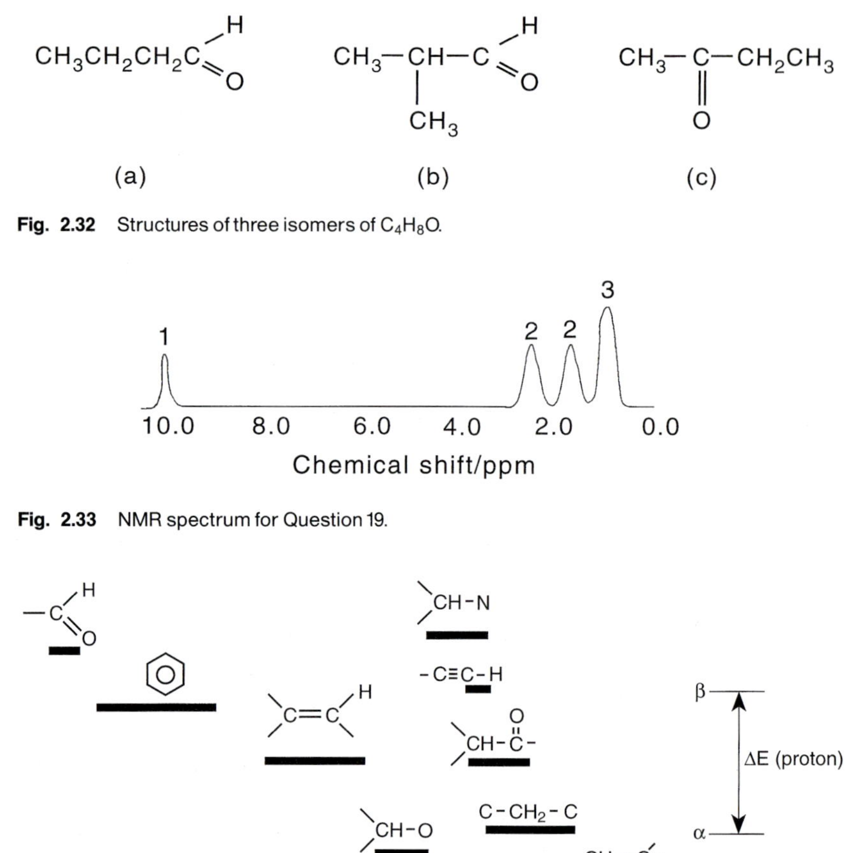

Fig. 2.32 Structures of three isomers of C_4H_8O.

Fig. 2.33 NMR spectrum for Question 19.

^{1}H NMR spectroscopy involves the use of the radio frequency electromagnetic waves to induce transitions of the nuclear spin of ^{1}H from "down" (α) to "up" (β).

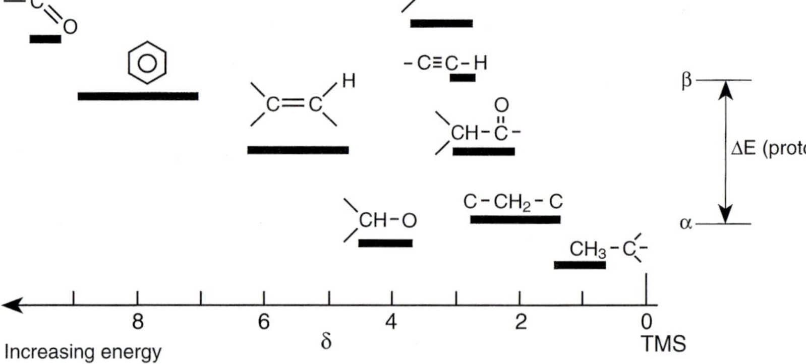

Fig. 2.34 Typical relative positions for different functional groups in a ^{1}H NMR spectrum. The energy difference between the α and β states is very small. The difference between the transition energies for different type of protons is even smaller. We usually use an energy scale relative to the standard tetramethylsilane, $Si(CH_3)_4$, (TMS):

$$\delta = \frac{\Delta E(\text{proton}) - \Delta E(\text{TMS proton})}{\Delta E(\text{TMS proton})} \times 10^6.$$

the relevant peak. Which isomer produced the spectrum? The data in Fig. 2.34 may be useful.

(b) Fig. 2.35 shows either the NMR spectrum of propanol, CH_3CH_2CHO, or of propanone, CH_3COCH_3. Which compound produced the spectrum? Explain the fine structure observed.

Answer

(a) Molecule A is the only isomer with proton ratios of 1 : 2 : 2 : 3. Examination of Fig. 2.33 shows that the chemical shift (δ) for the single

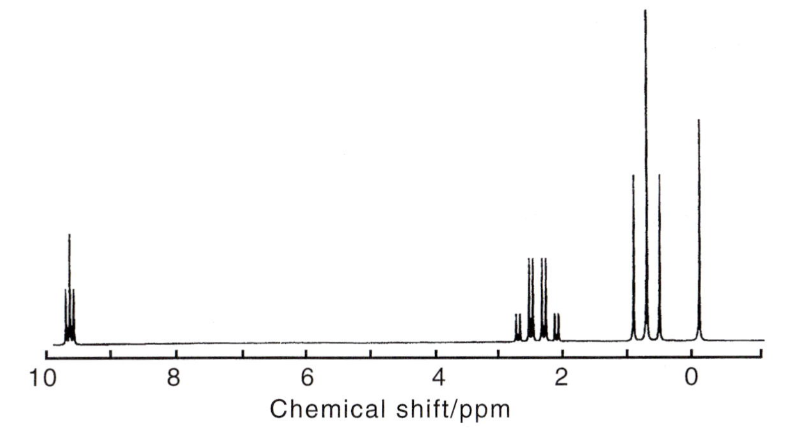

Fig. 2.35 The NMR spectrum of CH_3COCH_3 or CH_3CH_2CHO.

proton is at the value expected for a CHO proton, confirming this assignment.

(b) Fig. 2.35 shows the presence of a CHO group and C-H groups implying the molecule is CH_3CH_2CHO. The aldehyde proton near $\delta = 10$ is a triplet of intensities 1 : 2 : 1 resulting from coupling to the adjacent CH_2 group which contains two equivalent protons. The resonance near $\delta = 1$ corresponds to the CH_3 group which also couples to the CH_2 group producing a 1 : 2 : 1 triplet. Last, the signal between $\delta = 2$ and 3 corresponds to the CH_2 protons which couple to both the CHO and the CH_3 protons. The latter give a 1 : 3 : 3 : 1 quartet which is further split into doublets by the single CHO proton.

Question 20

Fig. 2.36 shows the proton NMR spectrum of a compound $C_{10}H_{12}O_2$. There is a strong absorption band in the IR spectrum at 1735 cm^{-1}, and a parent peak at m/e = 91 in its mass spectrum. Deduce a structural formula for the compound. The data given in Table 2.1 are useful.

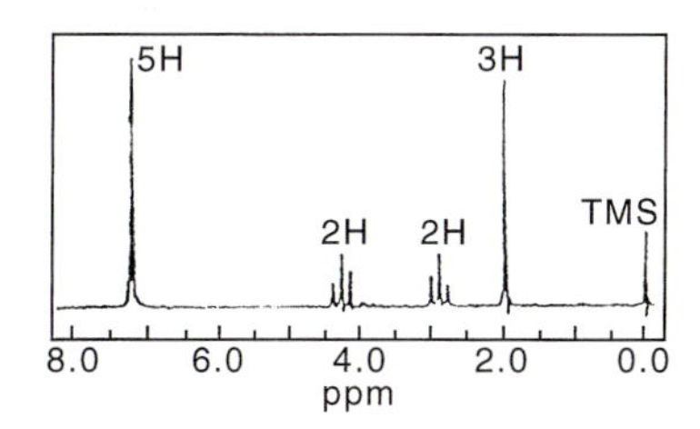

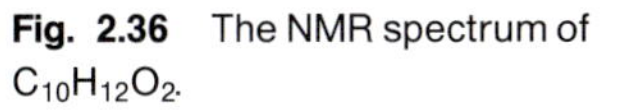

Fig. 2.36 The NMR spectrum of $C_{10}H_{12}O_2$.

Answer

The wavenumber of the infrared band together with the stoichiometry of the compound hints at the presence of an ester. The proton NMR implies four types of proton with an intensity ratio of 5 : 2 : 2 : 3. A likely structure is

$$C_6H_5CH_2CH_2CO_2CH_3.$$

Note that this can be distinguished from the isomer

$$C_6H_5CO_2CH_2CH_2CH_3$$

since the CH_3 protons do not couple to the CH_2 group(s).

3 Chemical energetics

3.1 Aims

The problems and answers in this chapter seek first to illustrate and develop the following core topics.

- Hess' law.
- Enthalpies of formation.
- Bond enthalpies.
- Lattice energies and the stability of solids.
- Enthalpy and entropy changes accompanying reactions.
- Prediction of the direction of chemical change via free energy changes.

Second, we aim to introduce several more sophisticated topics:

- the temperature variation of enthalpy changes,
- heat changes at constant pressure as compared to constant volume,
- the Second Law of Thermodynamics and the prediction of the direction of chemical changes,
- Ellingham diagrams, and
- the Born-Landé equation for the prediction of lattice energies.

3.2 Problems on core topics

Question 1

(a) What do you understand by the terms standard enthalpy of formation, ΔH_f°, and average bond enthalpy? State Hess' law.

(b) The average OH bond enthalpy in H_2O is 464 kJ mol^{-1}, yet the standard enthalpy of dissociation for the reaction

$$H_2O(g) \rightarrow OH(g) + H(g)$$

is +498 kJ mol^{-1}. Comment.

(c) The successive bond dissociation enthalpies of the C-H bonds in methane, CH_4, are 425, 445, 445 and 390 kJ mol^{-1} respectively. Calculate the average C-H bond enthalpy for methane.

(d) Calculate the enthalpy change accompanying the reaction

$$C_2H_2(g) + 2H_2(g) \rightarrow C_2H_6(g)$$

given that the enthalpies of formation of C_2H_2 and C_2H_6 are, respectively, +227 and −85 kJ mol^{-1}.

Answer

(a) The standard enthalpy of formation of a compound, ΔH_f°, is the enthalpy change involved in the formation of one mole of the substance from its elements, each element being in its usual form at one atmosphere pressure and the specified temperature.

The bond enthalpy is the enthalpy for the breaking of one mole of bonds. In the case of a diatomic molecule, this is the enthalpy change for the dissociation

$$AZ(g) \rightarrow A(g) + Z(g).$$

For polyatomic molecules of the type AZ_n, the average bond enthalpy is $\frac{\Delta H^\circ}{n}$ where ΔH° refers to the reaction

$$AZ_n(g) \rightarrow A(g) + nZ(g).$$

For polyatomics containing more than one type of bond, average bond enthalpies are estimated from the enthalpies of dissociation of several compounds containing bond types in common. For example knowledge of the enthalpy of dissociation of CH_4 and C_2H_6 permits average bond enthalpies to be found for C-C and C-H bonds.

The deduction of bond enthalpies for C-C and C-H from the enthalpy of dissociation of CH_4 and C_2H_6 is discussed on page 37 of C. P. Lawrence, A. Rodger and R. G. Compton, "Foundations of physical chemistry", (OCP 40).

Hess' law states that if a reaction is conducted in more than one stage, the overall enthalpy change is the sum of the enthalpy changes involved in the separate stages.

(b) The average bond enthalpy of the OH bond in H_2O is one half the enthalpy of the reaction

$$H_2O(g) \rightarrow O(g) + 2H(g)$$

As this value is 464 kJ mol^{-1}, and as the enthalpy of the reaction

$$H_2O(g) \rightarrow OH(g) + H(g)$$

is 498 kJ mol^{-1}, we conclude that for the reaction

$$OH(g) \rightarrow O(g) + H(g)$$

the enthalpy change is

$$\Delta H = 2 \times 464 - 498$$
$$= 430 \text{ kJ mol}^{-1}$$

(c) For the reaction

$$CH_4(g) \rightarrow C(g) + 4H(g)$$

the enthalpy change is

$$\Delta H = 425 + 445 + 445 + 390$$
$$= 1705 \text{ kJ mol}^{-1}$$

The average bond enthalpy is one quarter of this value: 426 kJ mol^{-1}

(d) To evaluate the sought enthalpy change we construct the following Hess cycle.

It should be noted that the First Law of Thermodynamics—that energy (specifically enthalpy) is conserved in chemical reaction—is assumed to be true in this and all subsequent Hess cycles.

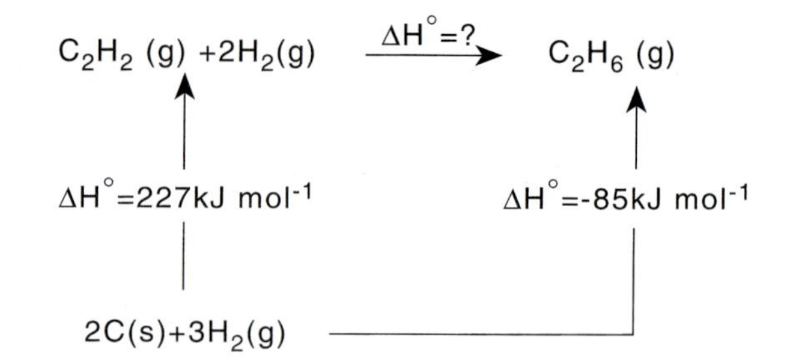

Applying Hess' law, the sought enthalpy quantity is

$$\Delta H^\circ = -227 - 85$$
$$= -312 \text{ kJ mol}^{-1}$$

Question 2

(a) Estimate the enthalpy of hydrogenation of cyclohexene

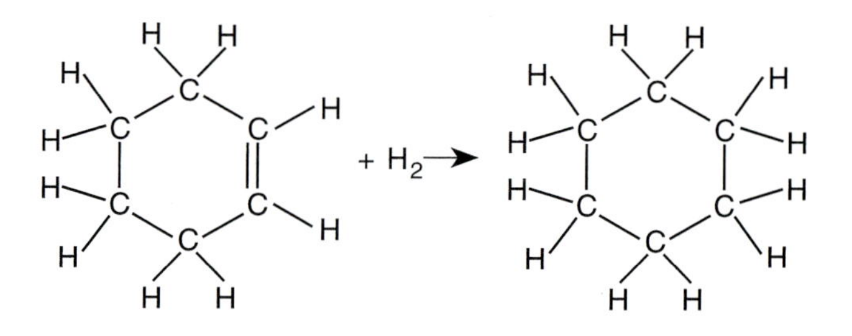

and that of cyclohexadiene,

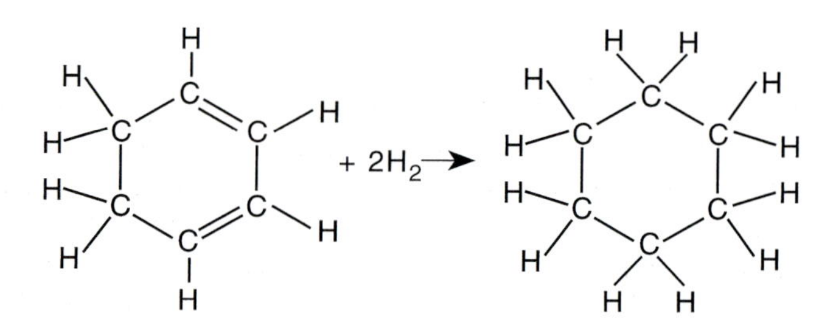

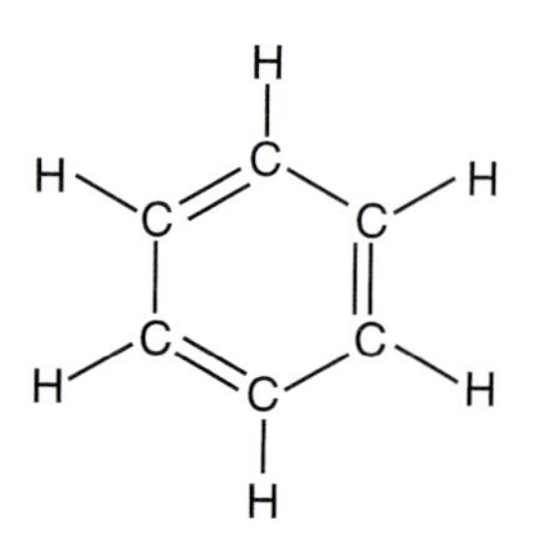

given the following average bond enthalpies measured in kJ mol^{-1}: D(C-H), 416; D(C-C), 346; D(C=C), 609; D(H-H), 436.

(b) Estimate the enthalpy of hydrogenation of benzene, C_6H_6, assuming that its structure is as shown in the margin.
The experimentally observed value is close to 210 kJ mol^{-1}. Comment.

Answer

(a) For the case of cyclohexene we adopt the following Hess cycle.

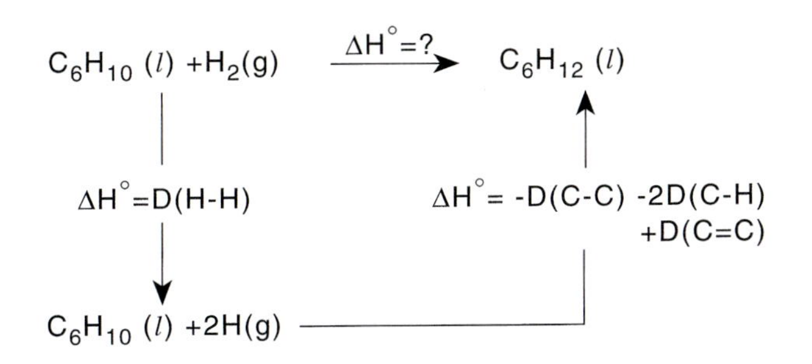

It follows that the required enthalpy change is

$$\begin{aligned}\Delta H^\circ &= -D(\text{C-C}) - 2D(\text{C-H}) + D(\text{C=C}) + D(\text{H-H})\\ &= -346 - 2 \times 416 + 609 + 436\\ &= -133\ \text{kJ mol}^{-1}\end{aligned}$$

Turning next to cyclohexadiene we use a similar Hess cycle.

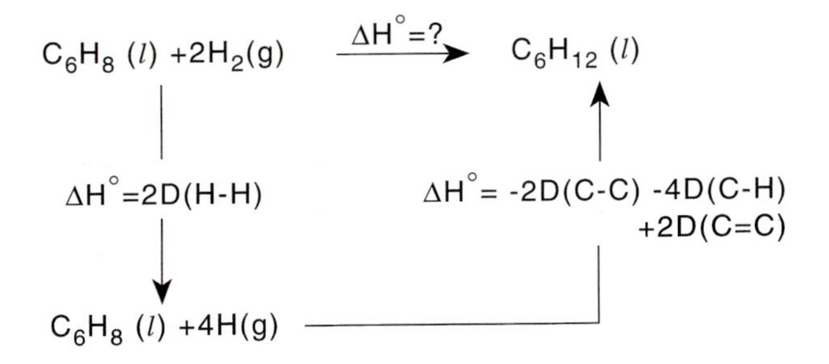

From this it may be deduced that the unknown is

$$\begin{aligned}\Delta H^\circ &= -2D(\text{C-C}) - 4D(\text{C-H}) + 2D(\text{C=C}) + 2D(\text{H-H})\\ &= -2 \times 346 - 4 \times 416 + 2 \times 609 + 2 \times 436\\ &= -266\ \text{kJ mol}^{-1}\end{aligned}$$

Looking at the two results above note that the enthalpy change for the hydrogenation of the two double bonds in cyclohexadiene is exactly twice that of the hydrogenation of the one double bond in cyclohexene.

(b) If the structure of benzene is assumed to be that drawn (1,3,5-cyclohexatriene) then the enthalpy of hydrogenation, ΔH°, of the three double bonds will be three times that of the one double bond in cyclohexene. It follows that,

$$\begin{aligned}\Delta H^\circ &= -3 \times 133\\ &= -399\ \text{kJ mol}^{-1}\end{aligned}$$

The experimental value is -210 kJ mol^{-1} so that the true structure of benzene must be more stable—by an amount corresponding to -189 kJ

mol^{-1}—than the structure drawn in the question. This shows it having three single and three double bonds. In reality all the carbon–carbon bonds in benzene have the same length as a result of the delocalisation of the electrons in the ring. This can be represented either in valence bond terms as resonance between several structures, including

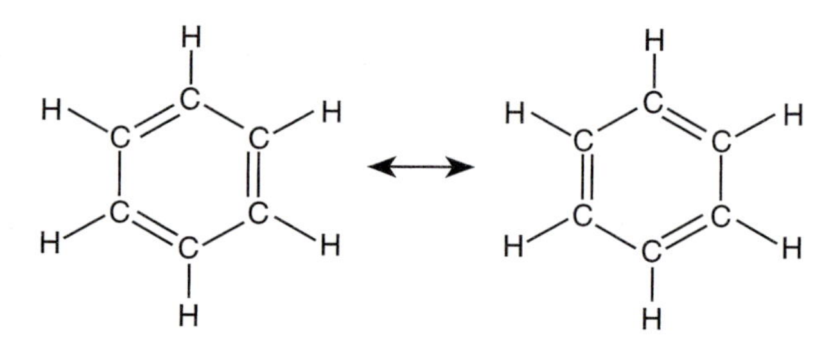

or in molecular orbital terms as orbitals extending over all the atoms in the ring.

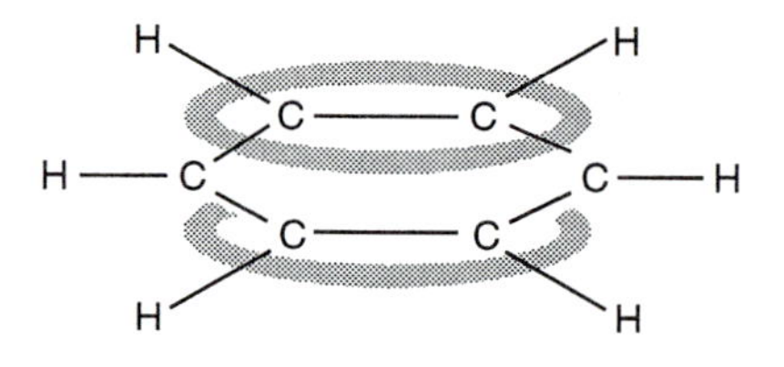

Question 3

Assuming the following enthalpies of combustion,

$$C(s) + O_2(g) \rightarrow CO_2(g) \qquad \Delta H^\circ = -393 \text{ kJ mol}^{-1}$$

$$H_2(g) + \frac{1}{2}O_2(g) \rightarrow H_2O(\ell) \qquad \Delta H^\circ = -286 \text{ kJ mol}^{-1}$$

calculate the standard enthalpies of formation of ethanol given that its enthalpy of combustion has the value of -1367 kJ mol^{-1}.

Answer

We use the Hess cycle below.

$C_2H_5OH(l) + 3O_2(g) \xrightarrow{\Delta H^\circ = -1367 \text{kJ mol}^{-1}} 2CO_2(g) + 3H_2O(l)$

$\uparrow$ $\Delta H_f^\circ = ?$ $\qquad\qquad$ $\uparrow$ $\Delta H^\circ = -2 \times 393 - 3 \times 286$ kJ mol^{-1}

$2C(s) + 3H_2(g) + \frac{1}{2}O_2(g) + 3O_2(g)$

From the cycle we see that

$$\Delta H_f^\circ = -2 \times 393 - 3 \times 286 + 1367$$
$$= -277 \text{ kJ mol}^{-1}$$

Question 4

Calculate the standard enthalpy of formation of propane gas from the following enthalpies of combustion:

$$C_3H_8(g) + 5O_2(g) \rightarrow 3CO_2(g) + 4H_2O(\ell) \qquad \Delta H^\circ = -2220 \text{ kJ mol}^{-1}$$
$$C(s) + O_2(g) \rightarrow CO_2(g) \qquad \Delta H^\circ = -393 \text{ kJ mol}^{-1}$$
$$H_2(g) + \frac{1}{2}O_2(g) \rightarrow H_2O(\ell) \qquad \Delta H^\circ = -286 \text{ kJ mol}^{-1}$$

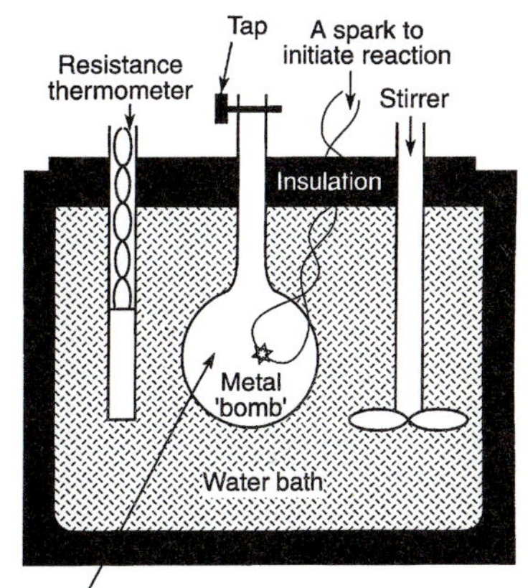

Fig. 3.1 An adiabatic bomb calorimeter. In measuring enthalpies of combustion, a small amount of liquid water is added to the bomb so that the further water generated via combustion is formed as a liquid corresponding to the standard state of water.

Answer

The following Hess cycle is helpful.

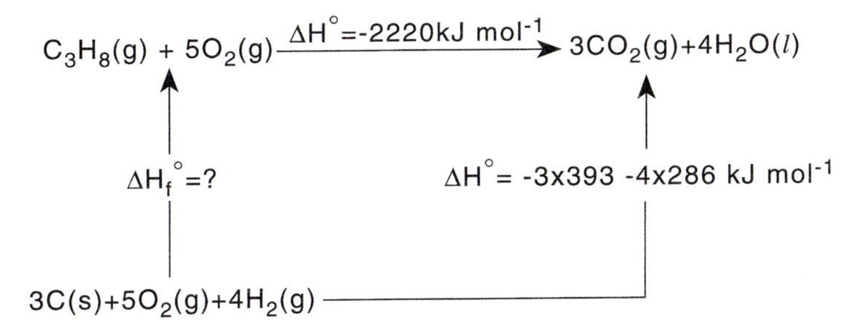

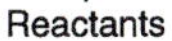

On inspecting the cycle it can be seen that

$$\Delta H_f^\circ = -3 \times 393 - 4 \times 286 + 2220$$
$$= -103 \text{ kJ mol}^{-1}$$

Question 5

In the combustion of a sample of 1.26 g of glucose ($C_6H_{12}O_6$) in an adiabatic bomb calorimeter (see Fig. 3.1), the temperature rose from 22.17 °C to 28.32 °C. The thermal heat capacity of the system was 3.185 kJ K^{-1}. Determine the enthalpy of combustion of glucose and its enthalpy of formation at 298 K given that the standard enthalpies of formation of $CO_2(g)$ and $H_2O(\ell)$ at 298 K are −393 and −286 kJ mol^{-1} respectively. [Relative atomic masses: C,12; H,1; O,16]

The word "adiabatic" implies that no heat is lost to the surroundings. All the heat released in the reaction is used to warm up the bomb and its contents.

Answer

The starting point is to calculate the number of moles of glucose combusted. The relative molecular mass of glucose is,

$$6 \times 12 + 12 \times 1 + 6 \times 16 = 180$$

Accordingly 1.26 g of glucose contains 1.26 / 180 moles (7×10^{-3} moles).

Next, if the thermal capacity of the system is -3.185 kJ K^{-1} then a temperature rise of 6.15 K must have been induced by the release of 6.15×3.185 kJ of heat. It follows that had one mole been combusted a quantity $(6.15 \times 3.185) / (7 \times 10^{-3})$ kJ of heat would have been released. So the enthalpy of combustion of glucose is -2798 kJ mol^{-1}.

To find the enthalpy of formation of glucose a Hess cycle is required.

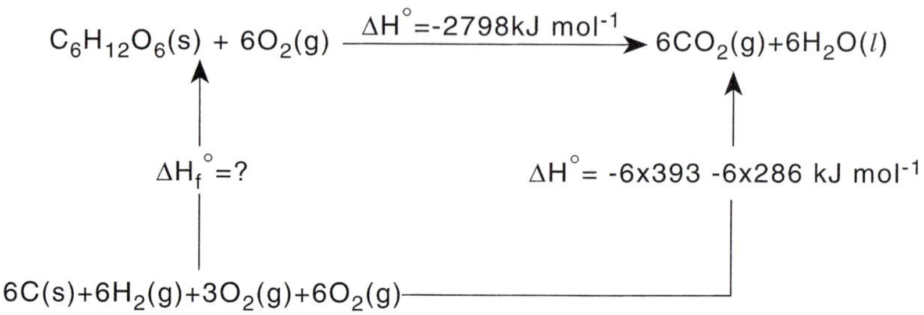

$$C_6H_{12}O_6(s) + 6O_2(g) \xrightarrow{\Delta H^\circ = -2798 \text{kJ mol}^{-1}} 6CO_2(g) + 6H_2O(l)$$

$\Delta H_f^\circ = ?$ (from $6C(s)+6H_2(g)+3O_2(g)+6O_2(g)$ up to $C_6H_{12}O_6(s) + 6O_2(g)$)

$\Delta H^\circ = -6x393\ -6x286$ kJ mol^{-1} (from $6C(s)+6H_2(g)+3O_2(g)+6O_2(g)$ up to $6CO_2(g)+6H_2O(l)$)

$$6C(s)+6H_2(g)+3O_2(g)+6O_2(g)$$

The inference is,

$$\begin{aligned} \Delta H_f^\circ &= -6 \times 393 - 6 \times 286 + 2798 \\ &= -1276 \text{ kJ mol}^{-1} \end{aligned}$$

Question 6

(a) When 1.04 g of cyclopropane was burnt in excess oxygen in a bomb calorimeter, the temperature of the calorimeter rose by 3.69 K. The total heat capacity of the calorimeter and its contents was 14.01 kJ K^{-1}. Determine the enthalpy of combustion of cyclopropane.

(b) The standard enthalpies of formation of water and carbon dioxide are -286 and -393 kJ mol^{-1} respectively. Find the standard enthalpy of formation of cyclopropane.
[Relative atomic masses: C,12; H,1]

The vigilant reader may object that the adiabatic bomb calorimeter measures the heat released on combustion under conditions of constant volume rather than those of constant pressure required by the definition of enthalpy changes. This issue is addressed in Question 20.

Answer

(a) This question is similar to the last problem and maybe solved by a similar sequence of steps.

Step 1. The relative molecular mass of C_3H_6 is $3 \times 12 + 6 \times 1 = 42$. Therefore 1.04 g of cyclopropane is 1.04/42 moles (0.0248 moles).

Step 2. As the heat capacity of the calorimeter was 14.01 kJ K^{-1} a temperature rise of 3.69 K corresponds to 14.01×3.69 kJ of heat released (by burning of 0.0248 moles).

Step 3. The combustion of one mole of C_3H_6 would give rise to $(14.01 \times 3.69) / 0.0248$ kJ. The enthalpy of combustion is therefore -2088 kJ mol^{-1}

(b) Unsurprisingly a Hess cycle is required!

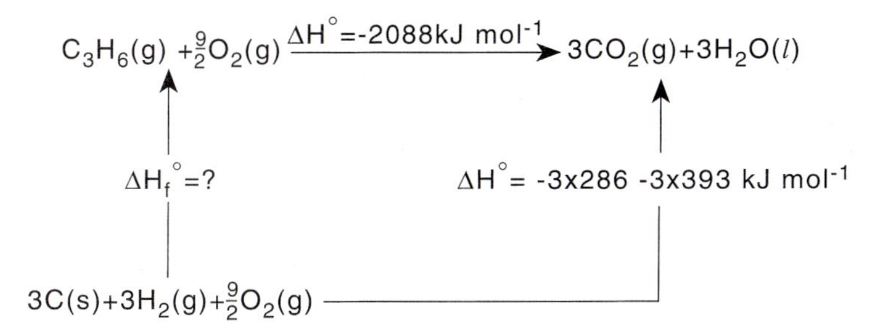

This shows that,

$$\Delta H_f^\circ = -3 \times 286 - 3 \times 393 + 2088$$
$$= +51 \text{ kJ mol}^{-1}.$$

Question 7

The standard enthalpy of combustion of graphite to form CO_2 is -393 kJ mol^{-1} and that of CO to form CO_2 is -283 kJ mol^{-1}. The standard enthalpies of dissociation of CO and O_2 are 1076 and 498 kJ mol^{-1} respectively. Use these data to evaluate the standard enthalpy of sublimation of graphite, ΔH°_{sub}. Why is it difficult to measure ΔH°_{sub} of graphite directly?

Answer

Construct the following Hess cycle:

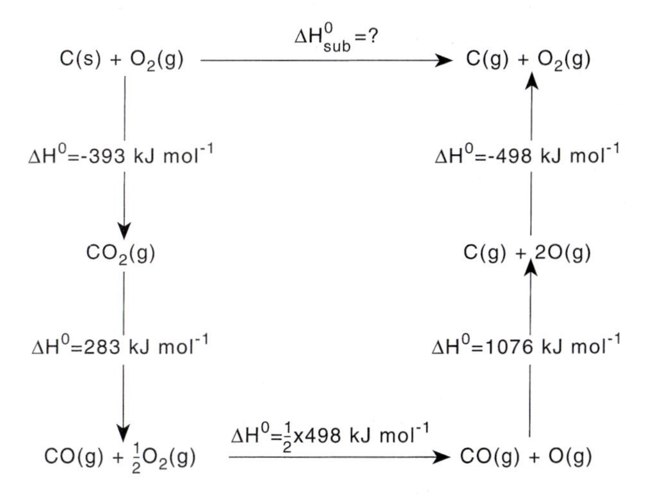

From the above it can be seen that

$$\Delta H^\circ_{sub} = -393 + 283 + \frac{1}{2} \times 498 + 1076 - 498$$
$$= 717 \text{ kJ mol}^{-1}$$

The direct measurement of this quantity is difficult for several reasons.

(i) The sublimation would need to take place in a vacuum or under an inert atmosphere otherwise CO and CO_2 would be formed by reaction.

(ii) The quantitative formation of gaseous carbon would be difficult to achieve.

(iii) The exclusive formation of monatomic carbon would be most improbable. In particular C_2 is a strongly bound molecule and would likely form.

The molecule C_2 is formed when a candle burns. Electronic transitions give the characteristic blue part of the candle flame.

Question 8

Cyclopropane is a gas whilst cyclohexane is a liquid at room temperature and pressure. Their structures are shown below.

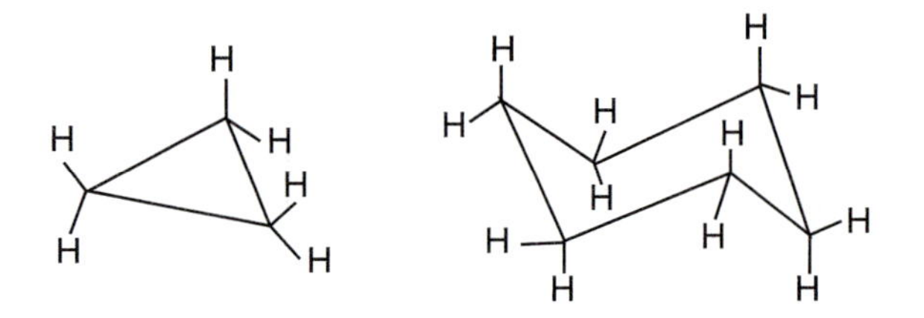

(a) Calculate the enthalpy change of combustion of cyclopropane and cyclohexane using all of the following data measured in kJ mol^{-1}.

Enthalpy change of combustion of carbon	$C(s) + O_2(g) \rightarrow CO_2(g)$	–393
Enthalpy change of combustion of hydrogen	$H_2(g) + \frac{1}{2} O_2 (g) \rightarrow H_2O(\ell)$	–286
Enthalpy change of sublimation of C	$C(s) \rightarrow C(g)$	716
Enthalpy change of dissociation of H_2 molecule	$H_2 (g) \rightarrow 2H(g)$	436
Enthalpy change of vaporisation of cyclohexane	$C_6H_{12} (\ell) \rightarrow C_6H_{12} (g)$	33
Average bond energy of C-C bond		346
Average bond energy of C-H bond		416

(b) What are the C-C-C bond angles in cyclopropane and cyclohexane. How do these compare with those expected for the molecule $C(CH_3)_4$?

(c) Compare the calculated values of the enthalpy change of combustion of cyclopropane and cyclohexane with the experimentally observed values which are respectively 2091 and 3920 kJ mol^{-1}. Comment on the result.

Answer

(a) For cyclopropane the following Hess cycle is used to find ΔH°_{comb} of this molecule.

C3H6(g) + 9/2 O2(g)
ΔH0comb=?
3CO2(g) + 3H2O(l)
ΔH0=3x346 + 6x416 kJ mol-1
ΔH0=-3x286 kJ mol-1
3C(g)+6H(g) + 9/2 O2(g)
3CO2(g)+3H2(g) + 3/2 O2(g)
ΔH0=-3x716 kJ mol-1
ΔH0=-3x393 kJ mol-1
ΔH0=-3x436 kJ mol-1
3C(s)+6H(g) + 9/2 O2(g)
3C(s)+3H2(g) + 9/2 O2(g)

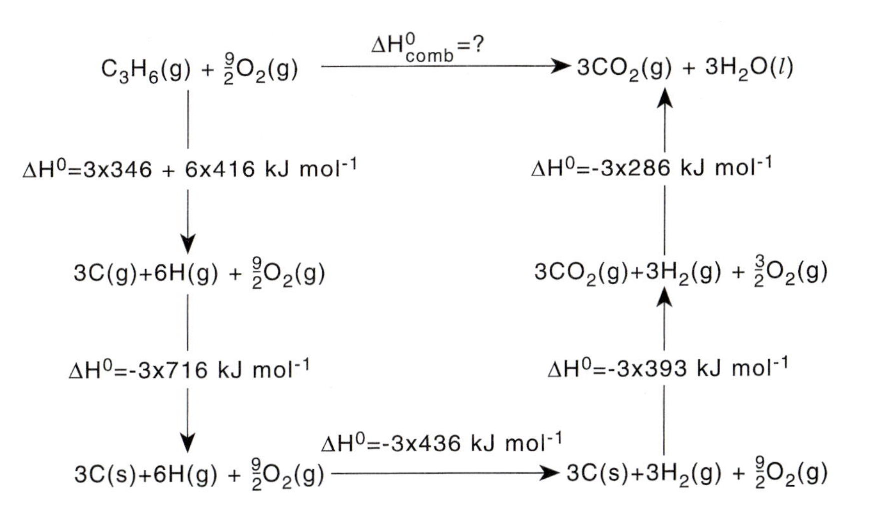

Summing the cycle gives

$$\Delta H^{\circ}_{comb} = 3 \times 346 + 6 \times 416 - 3 \times 716 - 3 \times 436 - 3 \times 393 - 3 \times 286$$
$$= -1959 \text{ kJ mol}^{-1}$$

A similar cycle can be drawn for cyclohexane but must additionally include the vaporisation of the liquid.

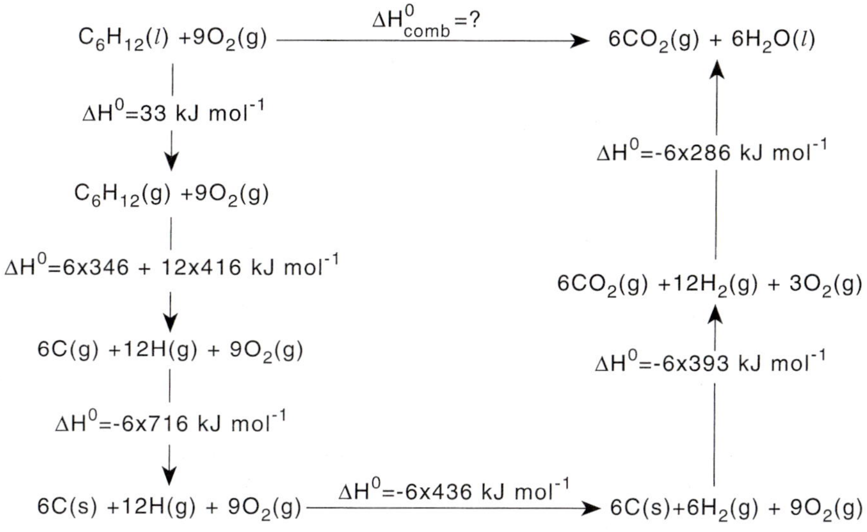

C6H12(l) +9O2(g)
ΔH0comb=?
6CO2(g) + 6H2O(l)
ΔH0=33 kJ mol-1
ΔH0=-6x286 kJ mol-1
C6H12(g) +9O2(g)
ΔH0=6x346 + 12x416 kJ mol-1
6CO2(g) +12H2(g) + 3O2(g)
6C(g) +12H(g) + 9O2(g)
ΔH0=-6x393 kJ mol-1
ΔH0=-6x716 kJ mol-1
6C(s) +12H(g) + 9O2(g)
ΔH0=-6x436 kJ mol-1
6C(s)+6H2(g) + 9O2(g)

In this case,

$$\Delta H^{\circ}_{comb} = 33 + 6 \times 346 + 12 \times 416 - 6 \times 716 - 6 \times 436 - 6 \times 393 - 6 \times 286$$
$$= -3885 \text{ kJ mol}^{-1}$$

(b) Inspection of the diagram of cyclopropane above shows that the three carbon atoms form an equilateral triangle so that the C-C-C bond angles are all 60°. This is significantly less than that seen in the molecule $C(CH_3)_4$ where the methyl groups are tetrahedrally distributed around the central carbon atom so that the C-C-C angles are all c.a. 109°. Cyclopropane is therefore a strained molecule.
The structure of cyclohexane shows that the six carbon atoms are not coplanar but rather the ring is puckered so that the bonds around each carbon atom are approximately tetrahedrally distributed. The arrangement—in which the overall shape of the molecule resembles a chair—removes any strained bond angles.

(c) For the case of cyclohexane the calculated enthalpy of combustion is less than 1% of the reported value corresponding to 55 kJ mol^{-1}. In contrast for the case of cyclopropane the discrepancy is as large as 132 kJ mol^{-1}. This reflects the strained nature of the bonds forming the cyclopropane and comparison with the 'average' values reflected in the bond enthalpies. In contrast the lack of strain in cyclohexane is evident from the closeness of values between the calculated and expected values.

Question 9

Using the following data calculate the electron affinity of F(g).

	ΔH°/kJ mol^{-1}
$Na(s) \rightarrow Na(g)$	107
$Na^+(g) + e^- \rightarrow Na(g)$	−502
$NaF(s) \rightarrow Na^+(g) + F^-(g)$	907
$Na(s) + \frac{1}{2}F_2(g) \rightarrow NaF(s)$	−574
$F_2(g) \rightarrow 2F(g)$	157

Answers

To find the electron affinity of F(g) it is necessary to construct a Born-Haber cycle. This is shown in Fig 3.2.
Summing the cycle shows that the

$$\begin{aligned}\text{electron affinity} &= -\tfrac{1}{2} \times 157 - 502 - 107 - 574 + 907 \\ &= -355 \text{ kJ mol}^{-1}.\end{aligned}$$

Question 10

Calcium reacts with fluorine to form fluorite, CaF_2 (Fig. 3.3).

Use the data given below to calculate the enthalpy change which occurs during the reaction

$$Ca(s) + F_2(g) \rightarrow CaF_2(s)$$

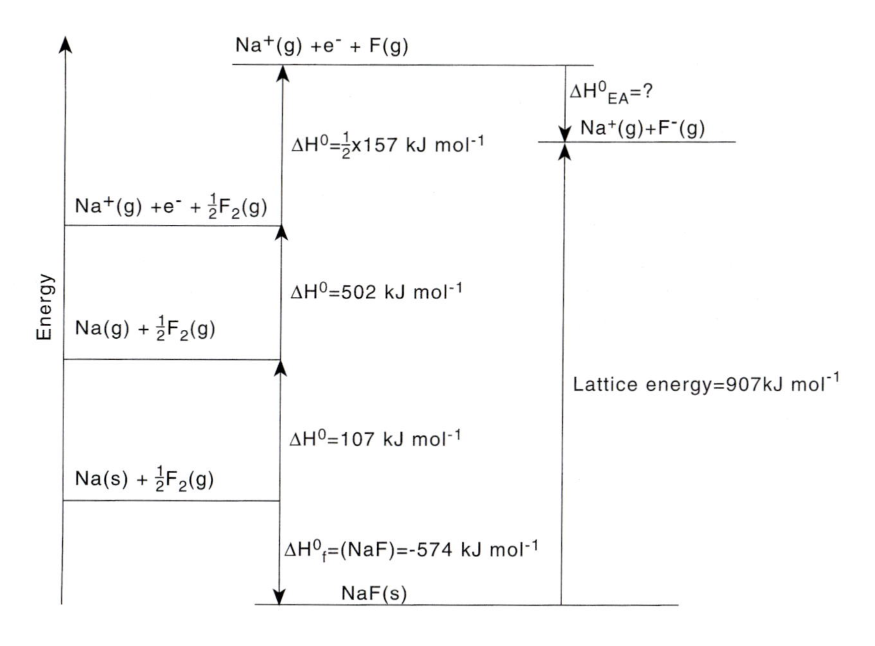

Fig. 3.2 The Born-Haber cycle for the formation of NaF.

	$\Delta H^\circ/\text{kJ mol}^{-1}$
$Ca^{2+}(g) + 2F^-(g) \rightarrow CaF_2(s)$	−2600
$Ca(s) \rightarrow Ca(g)$	178
$Ca(g) \rightarrow Ca^+(g) + e^-$	596
$Ca^+(g) \rightarrow Ca^{2+}(g) + e^-$	1152
$F_2(g) \rightarrow 2F(g)$	157
$F(g) + e^- \rightarrow F^-(g)$	−334

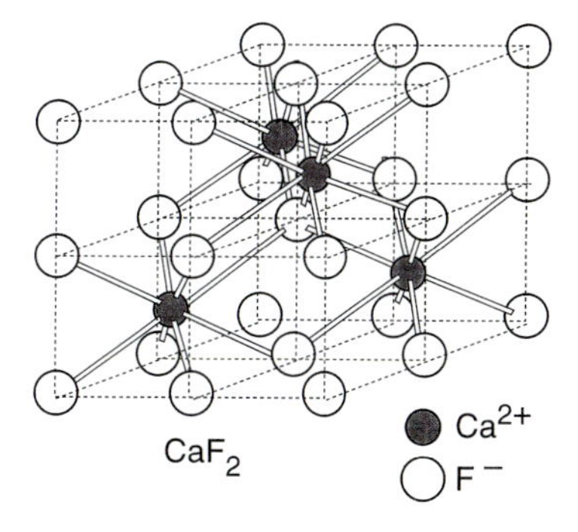

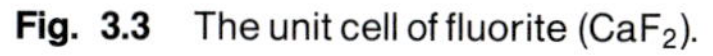
Fig. 3.3 The unit cell of fluorite (CaF_2).

Answer

In this question the sought quantity is the standard enthalpy of formation of CaF_2. Fig. 3.4 shows the Born-Haber cycle required. Applying Hess' law to the cycle shows

$$\Delta H_f^\circ = 178 + 596 + 1152 + 157 - 668 - 2600$$
$$= -1185 \text{ kJ mol}^{-1}.$$

Question 11

(a) Use the thermochemical data below to construct a Born-Haber cycle for the formation of NaCl.

Enthalpy of sublimation of solid sodium, $Na(s) \rightarrow Na(g)$ $\Delta H_{sub} = 107 \text{ kJ mol}^{-1}$

Enthalpy of dissociation of chlorine, $\frac{1}{2}Cl_2(g) \rightarrow Cl(g)$ $\Delta H_{diss} = 122 \text{ kJ mol}^{-1}$

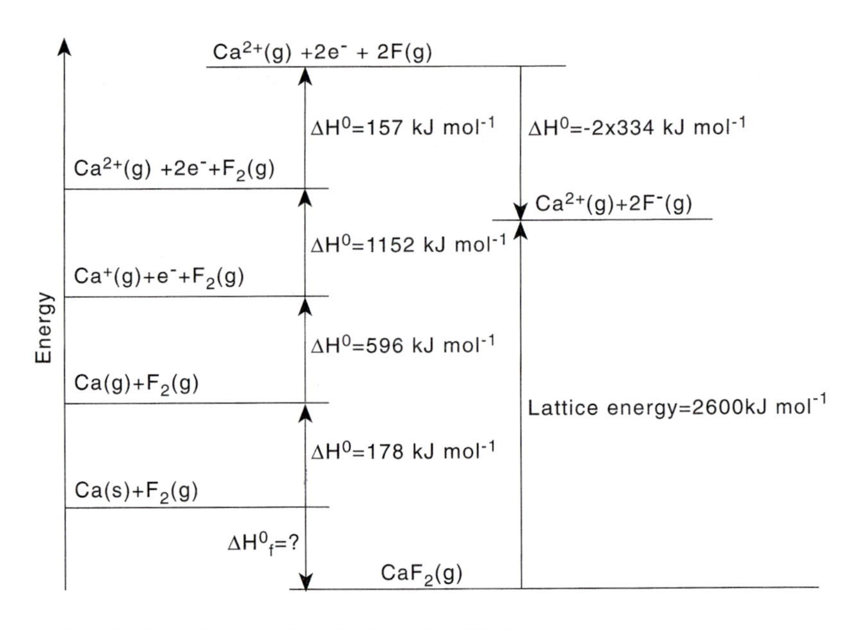

Fig. 3.4 The Born-Haber cycle for the formation of CaF_2.

First ionisation enthalpy of sodium, $Na(g) \rightarrow Na^+(g) + e^-$ $\quad \Delta H_{ion} = 502$ kJ mol^{-1}

Electron affinity of chlorine, $Cl(g) + e^- \rightarrow Cl^-(g)$ $\quad \Delta H_{EA} = -355$ kJ mol^{-1}

Enthalpy of formation of NaCl, $Na(s) + \frac{1}{2}Cl_2(g) \rightarrow NaCl(s)$ $\quad \Delta H_f = -411$ kJ mol^{-1}

(b) Using the additional data below, estimate the lattice energy of KCl.
Enthalpy of sublimation of solid potassium, $K(s) \rightarrow K(g)$ $\quad \Delta H_{sub} = 89$ kJ mol^{-1}
First ionisation enthalpy of potassium, $K(g) \rightarrow K^+(g) + e^-$ $\quad \Delta H_{ion} = 425$ kJ mol^{-1}
Enthalpy of formation of KCl, $K(s) + \frac{1}{2}Cl_2(g) \rightarrow KCl(s)$ $\quad \Delta H_f = -437$ kJ mol^{-1}

(c) Comment on the difference in the lattice enthalpies of NaCl and KCl.

(d) LiF, NaF and CsF all adopt the same crystal structure as NaCl (Fig. 3.5). Predict which solid has the (i) greatest and (ii) the smallest lattice energy.

(e) Using the following data, explain why the hypothetical compound $NaCl_2$ would be unstable with respect to NaCl and Cl_2, but $MgCl_2$ is the most stable chloride salt of magnesium.
Second ionisation enthalpy of Na, $Na^+(g) \rightarrow Na^{2+}(g) + e^-$ $\quad \Delta H_{ion} = 4569$ kJ mol^{-1}
First ionisation enthalpy of magnesium, $Mg(g) \rightarrow Mg^+(g) + e^-$ $\quad \Delta H_{ion} = 744$ kJ mol^{-1}
Second ionisation enthalpy of magnesium, $Mg^+(g) \rightarrow Mg^{2+}(g) + e^-$ $\quad \Delta H_{ion} = 1457$ kJ mol^{-1}

Answer

(a) The lattice energy of NaCl may be found by constructing a Born-Haber cycle analogous to that shown for NaF in Fig. 3.2. An alternative strategy is to 'add up' individual equations to find the sought transformation and its energy.

	$Na(s) \rightarrow Na(g)$	107 kJ mol^{-1}
+	$Na(g) \rightarrow Na^+(g) + e^-$	502 kJ mol^{-1}
+	$\frac{1}{2}Cl_2(g) \rightarrow Cl(g)$	122 kJ mol^{-1}
+	$Cl(g) + e- \rightarrow Cl^-(g)$	−355 kJ mol^{-1}
+	$NaCl(s) \rightarrow Na(s) + \frac{1}{2}Cl_2(g)$	411 kJ mol^{-1}
=	$NaCl(s) \rightarrow Na^+(g) + Cl^-(g)$	787 kJ mol^{-1}

The lattice energy of NaCl is 787 kJ mol^{-1}.

(b) Likewise for KCl

	$K(s) \rightarrow K(g)$	89 kJ mol^{-1}
+	$K(g) \rightarrow K^+(g) + e^-$	425 kJ mol^{-1}
+	$\frac{1}{2}Cl_2(s) \rightarrow Cl(g)$	122 kJ mol^{-1}
+	$Cl(g) + e^- \rightarrow Cl(g)$	−355 kJ mol^{-1}
+	$KCl(s) \rightarrow K^+(g) + Cl^-(g)$	437 kJ mol^{-1}
=	$KCl(s) \rightarrow K^+(g) + Cl^-(g)$	718 kJ mol^{-1}

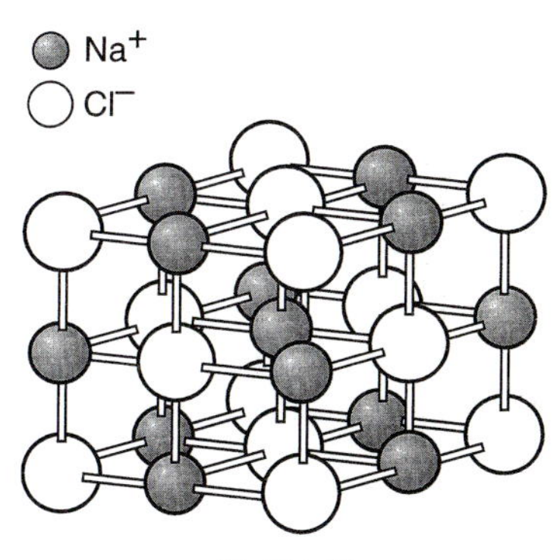

Fig. 3.5 The unit cell of rocksalt (sodium chloride).

(c) The NaCl lattice is more stable than the KCl lattice with respect to vaporisation of their constituent ions. This may be understood if we consider the packing of the ions in the two lattices. The rock salt structure shown in Fig. 3.5. Fig. 3.6 shows a two dimensional cross section from which it is clear that the ion–ion separation corresponds to the sum of the ionic radii of the cation and anion. The radii of these ions are: r(Na^+), 102pm; r(K^+), 138pm; r(Cl^-), 181pm. It is evident that the ions are closer together in NaCl than in KCl; the coulombic forces of attraction are greater so the lattice energy is correspondingly higher.

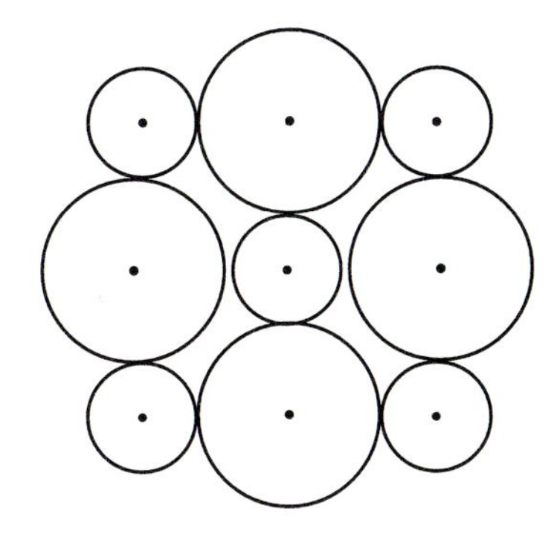

Fig. 3.6 Two dimensional cross-section view of NaCl.

(d) The most stable lattice out of LiF, NaF and CsF will be that which allows the ions to pack closest together: LiF. The least stable will be CsF. These predictions reflect the ionic radii which are as follows: r(Li^+), 76pm; r(Na^+), 102pm; r(Cs^+), 167pm.

(e) The energy required to remove a second electron from Na to form Na^{2+} *far* exceeds any of the energies included in the Born–Haber cycle for NaCl including the lattice energy. If $NaCl_2$ were formed the higher charge of the cation might lead to an increase in the lattice energy as a result of the coulombic forces of attraction but it is certainly the case that the increase from a value of 787 kJ mol^{-1} in NaCl would not be enhanced by a value in excess of the 4000–5000 kJ mol^{-1} required to offset the extra energy expenditure. In contrast the second ionisation energy of Mg^{2+} is much lower so that the extra lattice energy caused by the higher charge on the metal suffices to make the input of extra energy available.

Question 12

Predict, giving a reason in each case, whether the following reactions occur with a positive or negative change in entropy, $\Delta S°$.

(a) $NH_4Cl(s) \rightarrow NH_3(g) + HCl(g)$
(b) $2Na(s) + Cl_2(g) \rightarrow 2NaCl(s)$
(c) $NH_4NO_3(s) \rightarrow N_2O(g) + 2H_2O(g)$
(d) $2H_2(g) + O_2(g) \rightarrow 2H_2O(\ell)$
(e) $N_2O_4(g) \rightarrow 2NO_2(g)$
(f) $PH_3(g) + HI(g) \rightarrow PH_4I(s)$
(g) $2H_2O_2(aq) \rightarrow 2H_2O(\ell) + O_2(g)$
(h) $CO_2(g) + C(s) \rightarrow 2CO(g)$
(i) $3O_2(g) \rightarrow 2O_3(g)$
(j) $C_2H_4(g) + H_2(g) \rightarrow C_2H_6(g)$
(k) $Zn(s) + 2H^+(aq) \rightarrow Zn^{2+}(aq) + H_2(g)$

Answer

(a) One mole of solid NH_4Cl becomes two moles of gas (HCl and NH_3). Gases are *much* more disordered than solids as the molecules move about randomly throughout the full volume of the container as opposed to being fixed at lattice sites in a solid, so $\Delta S^\circ > 0$.

(b) In the reaction of solid sodium with gaseous chlorine to form solid sodium chloride the loss of the entropy associated with the disorder of the gas will ensure $\Delta S^\circ < 0$.

(c) The thermal decomposition of ammonium nitrate produces water vapour which, as a gas, has a much higher entropy than the solid ammonium nitrate so that $\Delta S^\circ > 0$.

(d) In the formation of liquid water from gaseous hydrogen and oxygen three moles of gas become one mole of liquid with a consequent major loss of disorder: $\Delta S^\circ < 0$.

(e) The dissociation of N_2O_4 to form NO_2 molecules is an entirely gas phase reaction. However, there is an increase in the number of molecules in moving from the reactants to the products so $\Delta S^\circ > 0$.

(f) In this reaction two moles of gas react to form one mole of solid PH_4I. There is a major loss of entropy so that $\Delta S^\circ < 0$.

(g) The decomposition of hydrogen peroxide in aqueous solution forms gaseous oxygen so that $\Delta S^\circ > 0$ as gases are more disordered than liquids.

(h) In this reaction of CO_2 with C, one mole of gas reacts with a solid to form two moles of gas. There is therefore a net entropy increase: $\Delta S^\circ > 0$.

(i) The allotropic conversion of dioxygen to form ozone molecules means there is a conversion of three moles of gas into two moles of gas, so that $\Delta S^\circ < 0$.

(j) In the hydrogenation of ethene to form ethane all the reactants and products are gaseous. However the reaction shows the conversion of two moles of reactant into one mole of products so that there is a net loss of disorder when the reaction occurs: $\Delta S^\circ < 0$.

(k) Simplistically one might expect $\Delta S^\circ > 0$ because one mole of gas is formed. However, the hydration of the Zn^{2+} ion is substantial and outweighs that of the two H^+ ions. Order is introduced into the system by virtue of the immobilisation of the water molecules close to the Zn^{2+} ion with the result that $\Delta S^\circ < 0$.

Question 13

(a) Explain why potassium iodide dissolves into water spontaneously, even though the process is endothermic.

(b) The following reaction in aqueous solution at 298 K lies strongly in favour of the products.

$$[Ni(NH_3)_6]^{2+}(aq) + 3\ en(aq) \rightarrow [Ni(en)_3]^{2+}(aq) + 6NH_3(aq)$$
$$(en = 1,2\text{-diaminoethane, } H_2NCH_2CH_2NH_2)$$

Fig. 3.7 shows the structure of the nickel complex.
The standard enthalpy and entropy changes accompanying the reaction are $\Delta H^\circ = -12\ kJ\ mol^{-1}$ and $\Delta S^\circ = +84\ J\ K^{-1}mol^{-1}$. Explain why the reaction lies in favour of the products and whether enthalpic or entropic factors are the most significant.

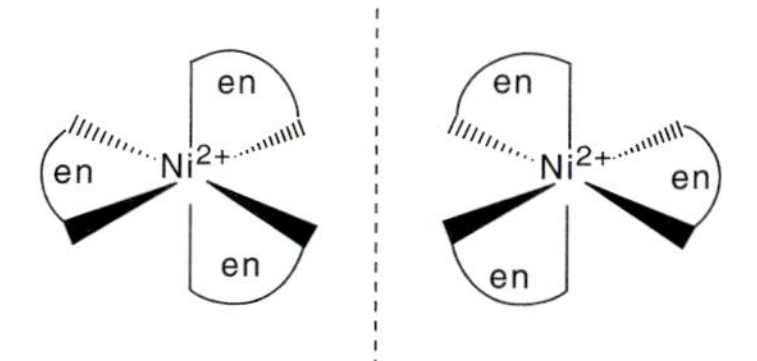

Fig. 3.7 The structure of the nickel complex, $Ni(en)_3^{2+}$ which exists as two enantiomers (non-superimposable mirror images).

Answers

(a) The reaction is

$$KI(s) + H_2O(\ell) \rightarrow K^+(aq) + I^-(aq)$$

for which $\Delta H^\circ > 0$ largely as a result of the fact that the iodide ions are poorly solvated due to their large size. The reaction is spontaneous as it is *entropy driven*; the much greater disorder of the K^+ and I^- ions when dissolved in solution as compared to when in the solid KI lattice energy ensures that $\Delta S^\circ > 0$.

(b) The ligand displacement reaction in which six moles of NH_3 are displaced by three chelating ligands, en, has a ΔH° value close to zero. However, the entropy change is large and positive and it is this fact which ensures that the equilibrium lies in favour of the product. This large and positive value of ΔS° arises since in the course of the reaction there is a net increase in the number of molecules in solution and hence in the disorder of the latter. The relative contribution of ΔH° and ΔS° can be found using the equation

$$\Delta G^\circ = \Delta H^\circ - T\Delta S^\circ$$

where ΔG° is the Gibbs free energy change for the reaction.
The enthalpic contribution to ΔG° is $-12\ kJ\ mol^{-1}$ whereas the $T\Delta S^\circ$ contribution is

$$(298 \times 84 \times 10^{-3}) = 25\ kJ\ mol^{-1} \text{ at } 298\ K.$$

Question 14

Phosphorus forms two chlorides, PCl_3 and PCl_5, whose standard enthalpies formation in the gaseous phase are -320 and $-375\ kJ\ mol^{-1}$.

(a) Calculate the average P-Cl bond enthalpy in PCl_3.

(b) Calculate the reaction enthalpy for

$$PCl_5(g) \rightarrow PCl_3(g) + Cl_2(g)$$

(c) Use part (b), to obtain a second estimate of the average P-Cl bond enthalpy.
(d) Discuss reasons why the results in (a) and (c) might be different.
(e) Given that the average P-P bond enthalpy is estimated to be 198 kJ mol^{-1}, would the species be P_2Cl_4 be expected to be stable?
The following data will be useful.

White phosphorous exists as the tetramer, P_4, in the solid state. Other forms of phosphorous, red and black, exist as polymeric structures. The enthalpy change quoted for P(s) → P(g) comprises both sublimation and atomisation.

$$Cl_2(g) \rightarrow 2Cl(g) \qquad \Delta H^\circ = +243 \text{ kJ mol}^{-1}$$
$$P(s) \rightarrow P(g) \qquad \Delta H^\circ = +317 \text{ kJ mol}^{-1}$$

Answer

(a) We consider the following Hess cycle.

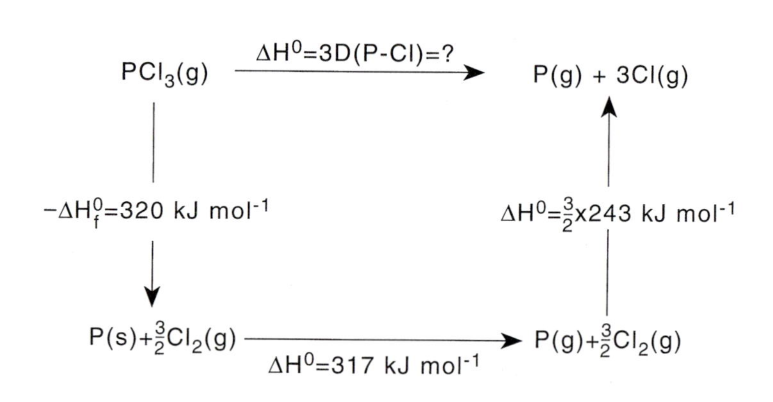

It follows that

$$\begin{aligned}\Delta H^\circ &= 3 \times D(P - Cl).\\ &= 320 + 317 + \tfrac{3}{2} \times 243\\ &= 1002 \text{ kJ mol}^{-1}\end{aligned}$$

so that D(PCl) = 334 kJ mol^{-1}

(b) The required Hess cycle is

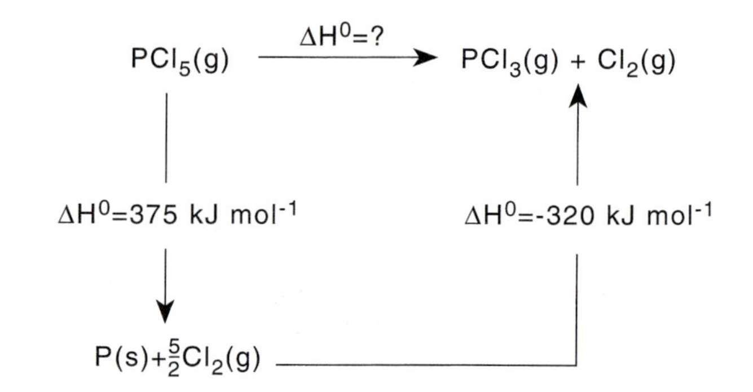

Therefore

$$\Delta H^\circ = 375 - 320$$
$$= 55 \text{ kJ mol}^{-1}$$

(c) Another simple Hess law is needed.

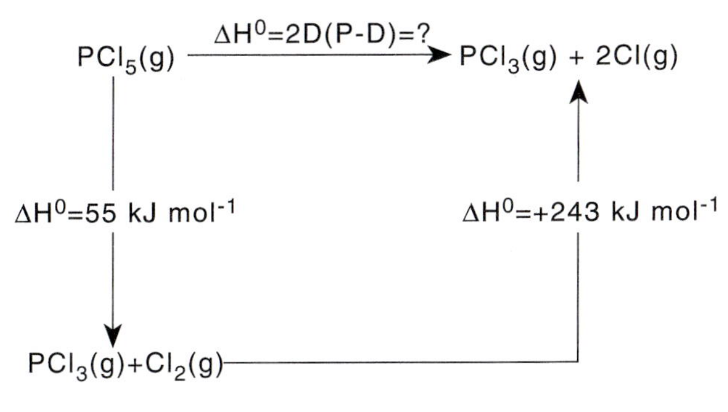

In this case,

$$2 \times D(P - Cl) = 55 + 243$$
$$= 298 \text{ kJ mol}^{-1}$$

It can be concluded that D(P-Cl) has a value of 149 kJ mol^{-1}. Taking a weighted average over the two bonds in PCl_5 and the three bonds in PCl_3 gives an bond energy of 260 kJ mol^{-1} for D(P-Cl)

(d) The value deduced in part (c) is significantly less than in part (a). Clearly the P-Cl bonds in PCl_3 are much stronger than the two weakest P-Cl bonds in PCl_5. It is notable that PCl_3 has a full octet of electrons in the outer shell where as the bonding in PCl_5 will involve "expansion of the octet" or, in other words, when a P atom combines with five Cl atoms to form PCl_5 some inner shell electrons must be promoted from the ground state configuration of [Ne] $3s^2$ $3p^3$ before five bonds can be formed.

(e) First we use the following cycle to ascertain if P_2Cl_4 is stable with respect to its elements.

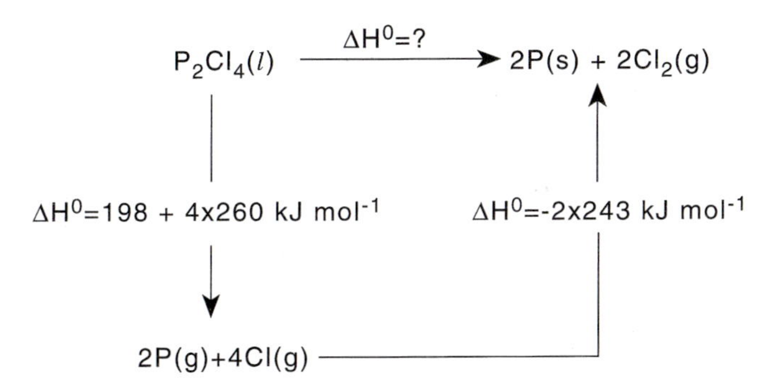

Notice in constructing the cycle an average value of 260 kJ mol^{-1} has been selected for D(P-Cl). From the cycle

$$\Delta H^\circ = 752 \text{ kJ mol}^{-1}.$$

This result shows that P_2Cl_4 is stable with respect to decomposition to its elements. However we also need to examine if decomposition into PCl_3 is thermodynamically possible. The following cycle is used.

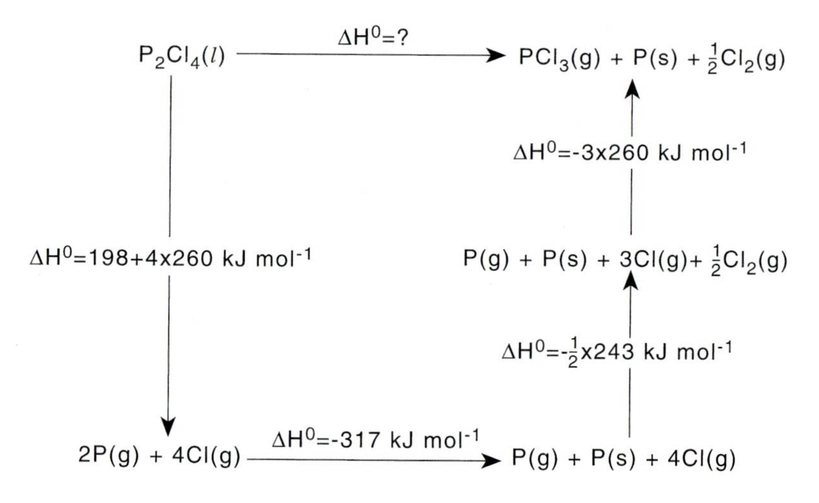

Summing the cycle gives

$$\Delta H^\circ = 20 \text{ kJ mol}^{-1}.$$

This indicates that the energetics are in the balance and, particularly given the uncertainty in the appropriate value of D(P-Cl) to use, the stability of P_2Cl_4 to decomposition into PCl_3 must be seen as an open question if judged in purely on enthalpic grounds. However the decomposition results in the formation of $1\frac{1}{2}$ moles of gas from the P_2Cl_4 and so is entropically very favourable. So in conclusion stability of P_2Cl_4 with respect to decomposition to PCl_3 is unlikely.

The values of S° quoted in Questions 15 and 18 are the absolute entropies of the substances specified.

The evaluation of absolute entropies requires knowledge of the Third Law of Thermodynamics which states that the entropy of a perfect crystal is zero at a temperature of zero kelvin. This corresponds therefore to perfect order. Assuming the third law, absolute entropies can be measured by calorimetric means as described by G. Price, " Thermodynamics of Chemical Processes", (OCP 56).

The three laws of thermodynamics may be summarised as follows.
(i) In an isolated system the internal energy is zero.
(ii) The entropy of the Universe is always increasing.
(iii) The entropy of a perfect crystal at zero kelvin is zero.

Question 15

(a) Given the standard enthalpies and absolute entropies below, calculate the free energy change accompanying the following reactions at 298 K.
(i) $2NO_2(g) \rightarrow 2NO(g) + O_2(g)$
(ii) $2NO_2(g) \rightarrow 2O_2(g) + N_2(g)$

Substance	ΔH_f° / kJ mol^{-1}	S° / J K^{-1} mol^{-1}
O_2	0	205
N_2	0	192
NO	90	211
NO_2	33	240

(b) Comment on the values deduced.

Answer

(a) (i) The first step is to calculate the entropy change for the reaction

$$2NO_2(g) \rightarrow 2NO(g) + O_2(g)$$

for which

$$\begin{aligned}\Delta S^\circ &= 2S^\circ(NO(g)) + S^\circ(O_2(g)) - 2S^\circ(NO_2(g))\\ &= 2 \times 211 + 205 - 2 \times 240\\ &= 147\ J\ K^{-1}\ mol^{-1}.\end{aligned}$$

The value reassuringly points to the expected entropy increase accompanying the conversion of two moles of gas into three moles of gas.

The second step requires the value of ΔH° for the same reaction. We lose the need to draw a Hess cycle by proceeding algebraically:

$$\begin{array}{llr} & N_2(g) + O_2(g) \rightarrow 2NO(g) & \Delta H^\circ = 2 \times 90 = 180\ kJ\ mol^{-1}\\ \text{minus} & N_2(g) + 2O_2(g) \rightarrow 2NO_2(g) & \Delta H^\circ = 2 \times 33 = 66\ kJ\ mol^{-1}\\ = & 2NO_2(g) \rightarrow 2NO(g) + O_2(g) & \Delta H^\circ = 180 - 66 = 114\ kJ\ mol^{-1}\end{array}$$

The final step utilises the equation

$$\Delta G^\circ = \Delta H^\circ - T\Delta S^\circ$$

so that for the reaction of interest

$$\begin{aligned}\Delta G^\circ &= 114 - 298 \times 147 \times 10^{-3}\\ &= 70\ kJ\ mol^{-1}.\end{aligned}$$

This shows that the favoured direction of reaction is that of NO reacting with O_2 to form NO_2.

(ii) For the reaction

$$2NO_2(g) \rightarrow 2O_2(g) + N_2(g)$$

the calculation of ΔG° proceeds through the steps which are exactly analogous to those in part (i).

Step 1 $\Delta S^\circ = 2S^\circ(O_2(g)) + S^\circ(N_2(g)) - 2S^\circ(NO_2(g))$

$$\begin{aligned}&= 2 \times 205 + 192 - 2 \times 240\\ &= 122\ J\ K^{-1}\ mol^{-1}.\end{aligned}$$

Step 2 The enthalpy of formation of NO_2 tells us directly that for the reaction

$$2NO_2(g) \rightarrow 2O_2(g) + N_2(g)$$

$$\begin{aligned}\Delta H^\circ &= -2 \times \Delta H_f^\circ(NO_2(g))\\ &= -66\ kJ\ mol^{-1}.\end{aligned}$$

Step 3 $\Delta G^\circ = \Delta H^\circ - T\Delta S^\circ$

$$\begin{aligned}&= -66 - 298 \times 122 \times 10^{-3}\\ &= -102\ kJ\ mol^{-1}.\end{aligned}$$

(b) Comparison of the data deduced in part (a) shows that NO_2 is stable with respect to decomposition into NO and O_2 but not to N_2 and O_2. The

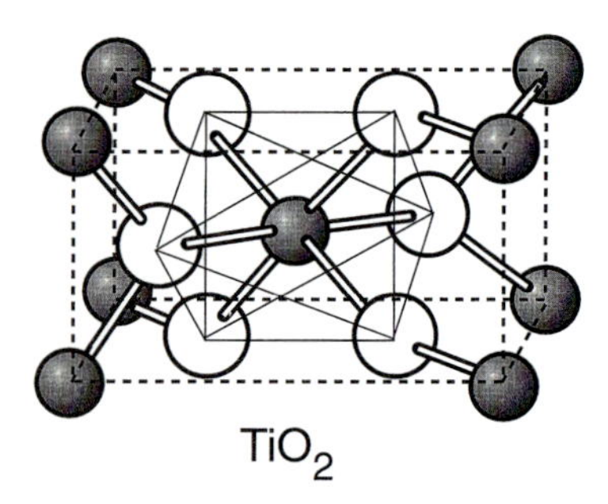

Fig. 3.8 The unit cell of rutile (TiO_2).

The value of $\Delta S°$ quoted refers to the temperature of 298 K. In calculating the answer to part (c), this value should be assumed to be temperature independent. In reality, this will not be strictly true, as the entropy of substances increase more strongly with temperature than, for example, do enthalpies. In the case of interest, the entropy of CO will likely increase *more* strongly with temperature than those of the solid species. So that in practice, any feasibility will be enhanced *above* the predictions in the answer.

difference reflects the very strong triple bond in N_2 which ensures the latter process is exothermic. In both cases $\Delta S°$ is positive and large. This is because both reactions involve two moles of gaseous reactant forming three moles of gaseous products.

Question 16

Titanium occurs naturally as the mineral rutile, TiO_2 (Fig. 3.8). One possible method of its extraction involves the reduction of the mineral by heating it with carbon:

$$TiO_2(s) + 2C(s) \rightarrow Ti(s) + 2CO(g)$$

(a) For this reaction, $\Delta S°$ has the value $+366\ J\ K^{-1}\ mol^{-1}$. Comment on the sign of this quantity.

(b) Calculate $\Delta H°$ for the reaction, given the standard enthalpies of formation of $TiO_2(s)$ and CO(g) are -944 and $-111\ kJ\ mol^{-1}$ respectively.

(c) Estimate ΔG for this reaction at 2200 K and comment on the feasibility of the reaction at this temperature.

Answer

(a) $\Delta S°$ is large and positive as the reaction involves three moles of solid being converted to two moles of gas (and one mole of solid).

(b) $\Delta H°$ for the reaction can be determined algebraically by writing the equation corresponding to the formation of CO_2 and H_2O from their constituent elements in their standard state:

$$2C(s) + O_2(g) \rightarrow 2CO(g) \quad \Delta H° = 2 \times \Delta H_f^\circ(CO)$$
$$= -2 \times 111$$
$$= -222\ kJ\ mol^{-1}$$
$$Ti(s) + O_2(g) \rightarrow TiO_2(s) \quad \Delta H° = \Delta H_f^\circ(TiO_2)$$
$$= -944\ kJ\ mol^{-1}$$

Subtracting gives

$$2C(s) + TiO_2(s) \rightarrow Ti(s) + O_2(g) + 2CO(g)$$

for which

$$\Delta H° = -222 - (-944)$$
$$= 722\ kJ\ mol^{-1}.$$

(c) The relevant equation is

$$\Delta G = \Delta H° - T\Delta S°$$
$$= 722 - 2200 \times 366 \times 10^{-3}$$
$$= -83\ kJ\ mol^{-1}$$

This shows that the reaction is favourable at the temperature specified. Note that the reaction is strongly endothermic but is entropy driven provided the temperature is high enough.

Question 17

Chemical processes move in directions in which the products have lower Gibbs free energy than the reactants. The relative contributions of the enthalpy and entropy changes accompanying the reaction are given by

$$\Delta G^\circ = \Delta H^\circ - T\Delta S^\circ$$

Comment on the following in the light of the above.

(a) Ice melts at atmospheric pressure only if the surrounding temperature rises above 0 °C.

(b) When sodium hydrogen carbonate is added to dilute hydrochloric acid, the temperature of the reaction mixture drops. Despite this, the reaction is spontaneous.

(c) Camping gas (liquid butane) combusts.

Answer

(a) The melting of ice is an endothermic process; energy has to be supplied to break the hydrogen bonds in the solid. However the solid to liquid change is accompanied by an increase in entropy as the liquid is more disordered than the solid. Looking at the equation

$$\Delta G^\circ = \Delta H^\circ - T\Delta S^\circ$$

and noting that ΔG° must be negative for a spontaneous change it can be seen that this will occur when

$$T \geq \frac{\Delta H^\circ}{\Delta S^\circ}$$

with the inequality corresponding to the melting point.

(b) The reaction is

$$HCO_3^-(aq) + H^+(aq) \rightarrow H_2O(\ell) + CO_2(g)$$

This is an endothermic process as the temperature is noted to drop on adding sodium hydrogen carbonate to HCl. Since the reaction proceeds spontaneously it must therefore be entropy driven. This can be appreciated once it is recognised that the reaction involves two solution (liquid) phase species forming gaseous CO_2 (and water).

(c) The reaction is

$$C_4H_{10}(\ell) + \frac{13}{2}O_2(g) \rightarrow 4CO_2(g) + 5H_2O(\ell)$$

This is evidently exothermic given the use of liquid butane in the camping gas context. However, ΔS° is strongly negative and unfavourable for the reaction as it involves the conversion of $\frac{13}{2}$ moles of oxygen into 4 moles of CO_2. It can be seen that ΔH° must outweigh the $T\Delta S^\circ$ term, at least at ambient temperatures.

Question 18

Chlorofluorocarbons (CFCs) have been widely used as refrigerants as well as aerosol propellants, blowing agents for plastic foams, and solvents. Dichlorodifluoromethane, CF_2Cl_2 (or CFC-12), is a typical member of this

Fluorinated hydrocarbons were developed in 1930 by General Motors Research laboratories as non-toxic, non-flammable refrigerants to replace SO_2 and NH_3 then in use.

class of compounds, some of which are now outlawed under international treaty. It has been suggested that sodium oxalate at elevated temperatures could be used to destroy existing stockpiles. One example of the proposed reaction, using CFC-12, is

$$CF_2Cl_2(g) + 2Na_2C_2O_4(s) \rightarrow 2NaF(s) + 2NaCl(s) + C(s) + 4CO_2(g)$$

Given the data below, is the reaction spontaneous at 298 K? Is there an upper limit to the temperature at which the reaction can be run?

CFCs are thought to damage the atmosphere through photolysis in the stratosphere:

$$CF_2Cl_2 + h\nu \rightarrow CF_2Cl + Cl.$$

The resulting chlorine atoms can destroy the ozone layer through catalytic chemistry such as the following.

$$Cl + O_3 \rightarrow ClO + Cl_2$$
$$ClO + O \rightarrow Cl + O_2$$

Substance	ΔG_f° / kJ mol^{-1}	ΔH_f° / kJ mol^{-1}	S° / J K^{-1} mol^{-1}
Sodium oxalate		−1318	150
Sodium			51
Carbon			6
Oxygen			205
CFC-12	−442		
Carbon dioxide (g)	−394		
Sodium fluoride	−544		
Sodium chloride	−384		

Answer

In order to quantify ΔG° for the reaction of interest the starting point is to focus on the formation of sodium oxalate from its component elements in their ground state:

$$2C(s) + 2O_2(g) + 2Na(s) \rightarrow Na_2C_2O_4(s)$$

for which ΔH_f° has a value of -1318 kJ mol^{-1}. It is also possible to find ΔS^0 for this reaction by way of the absolute entropies provided in the data.

$$\begin{aligned}\Delta S_f^\circ &= S^\circ(2Na_2C_2O_4(s)) - 2S^\circ(Na(s)) - 2S^\circ(O_2(g)) - 2S(C(s)) \\ &= 150 - 2 \times 51 - 2 \times 205 - 2 \times 6 \\ &= -374 \text{ J K}^{-1} \text{ mol}^{-1}\end{aligned}$$

The negative sign is not unexpected given the conversion of gases to solid in the reaction of interest. It follows that for the same reaction at 298 K

$$\begin{aligned}\Delta G_f^\circ &= \Delta H_f^\circ - T\Delta S_f^\circ \\ &= -1318 - 298 \times (-374 \times 10^{-3}) \\ &= -1207 \text{ kJ mol}^{-1}\end{aligned}$$

Consider the generalised Hess cycle below.

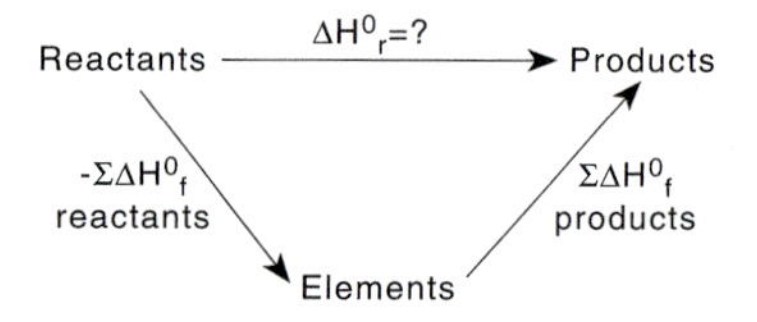

It follows that,

$$\Delta H_r^\circ = \sum_{\text{products}} \Delta H_f^\circ - \sum_{\text{reactants}} \Delta H_f^\circ.$$

Analogous expressions can be derived for ΔG_r° and ΔS_r°.
These results will be used in Questions 18 and 22.

Next consider the reaction of interest for which

$$\Delta G^\circ = \sum_{\text{products}} \Delta G_f^\circ - \sum_{\text{reactants}} \Delta G_f^\circ$$

$$\begin{aligned}\Delta G^\circ &= 2 \times \Delta G_f^\circ(NaF(s)) + 2 \times \Delta G_f^\circ(NaCl(s)) + \Delta G_f^\circ(C(s)) + 4 \\ &\quad \times \Delta G_f^\circ(CO_2(g)) - \Delta G_f^\circ(CF_2Cl_2(g)) - 2 \times \Delta G_f^\circ(2Na_2C_2O_4(s)) \\ &= -2 \times 544 - 2 \times 384 + 0 - 4 \times 394 + 442 + 2 \times 1207 \\ &= -576 \text{ kJ mol}^{-1}\end{aligned} \quad (3.1)$$

It can be concluded that the reaction is therefore viable at 298 K. However for the process to occur rapidly (with fast kinetics; chapter 4) it is necessary to consider how the thermodynamic feasibility varies with temperature since increasing temperature will enhance the rate of reaction.

Assuming $\Delta H_f^\circ(Na_2C_2O_4(s))$ and $\Delta S_f^\circ(Na_2C_2O_4(s))$ do not vary with temperature, then we can write

$$\Delta G_f^\circ(CF_2Cl_2(g)) = -1318 + T \times 374 \times 10^{-3}.$$

Substituting this equation into (3.1) we see that ΔG° for the reaction becomes zero when

$$\begin{aligned}\Delta G^\circ &= 2 \times \Delta G_f^\circ(NaF(s)) + 2 \times \Delta G_f^\circ(NaCl(s)) + \Delta G_f^\circ(C(s)) \\ &\quad + 4 \times \Delta G_f^\circ(CO_2(g)) - \Delta G_f^\circ(CF_2Cl_2(g)) - 2 \times \Delta G_f^\circ(NaC_2O_4(s)) \\ &= -2 \times 544 - 2 \times 384 + 0 - 4 \times 394 + 442 \\ &\quad - 2 \times (-1318 + T \times 374 \times 10^{-3}) \\ &= 0.\end{aligned}$$

It follows that at temperatures in excess of 473 K the process will be unfavourable.

3.3 Taking it further

Question 19: ΔH and temperature

It is frequently assumed that values of ΔH° are rather temperature insensitive. To question this, consider the standard enthalpy of formation of gaseous water, $H_2O(g)$, which at 298 K is -242 kJ mol^{-1}. Find the corresponding value at 600 K by using the following values of the molar heat capacities, C_p, at constant pressure: $H_2(g)$, 28.8 J K^{-1} mol^{-1}; $O_2(g)$, 29.4 J K^{-1} mol^{-1}; $H_2O(g)$, 33.6 J K^{-1} mol^{-1}. These values may be realistically assumed independent of temperature in the range 298 to 600 K.

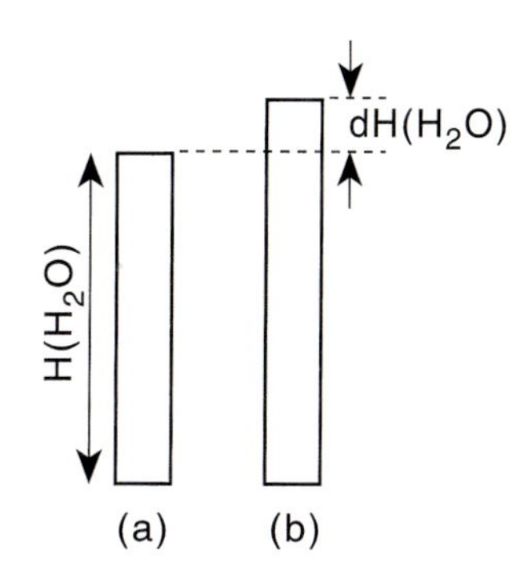

Fig. 3.9 Enthalpy of water (a) before and (b) after the addition of an amount of heat, dq.

Answer

By definition, the enthalpy changes are heat changes at constant pressure. Thus, if an amount, dq, is added to gaseous H_2O under conditions of constant pressure and produces an increase in temperature, dT, we can write:

$$dH^\circ(H_2O) = dq = C_p(H_2O)dT$$

where $H^\circ(H_2O)$is the enthalpy of water, and $dH^\circ(H_2O)$ is the increase in enthalpy of the water (Fig 3.9). It follows that

$$C_p(H_2O) = \frac{dH^\circ(H_2O)}{dT}$$

where in terms of calculus $\dfrac{dH^\circ(H_2O)}{dT}$ is the first derivative of $H^\circ(H_2O)$ with respect to T. Likewise,

$$C_p(H_2) = \frac{dH^\circ(H_2)}{dT} \quad \text{and} \quad C_p(O_2) = \frac{dH^\circ(O_2)}{dT}$$

We are interested in the temperature dependence of the standard enthalpy of formation of gaseous water defined as the enthalpy change accompanying the reaction

$$H_2(g) + \frac{1}{2}O_2(g) \rightarrow H_2O(g)$$

where
$$\Delta H_f^\circ = H^\circ(H_2O) - H^\circ(H_2) - \frac{1}{2}H^\circ(O_2)$$

Thus
$$\frac{d\Delta H_f^\circ}{dT} = \frac{dH^\circ(H_2O)}{dT} - \frac{dH^\circ(H_2)}{dT} - \frac{1}{2}\frac{dH^\circ(O_2)}{dT}$$
$$= C_p(H_2O) - C_p(H_2) - \tfrac{1}{2}C_p(O_2)$$
$$= 33.6 - 28.8 - \tfrac{1}{2} \times 29.4$$
$$= -9.9 \text{ J mol}^{-1}$$

It follows that

$$\frac{d\Delta H_f^\circ}{dT} = \frac{\Delta H_f^\circ(600) - \Delta H_f^\circ(298)}{600 - 298} = -9.9 \text{ J K}^{-1} \text{ mol}^{-1}$$

In general for any reaction, $\frac{d\Delta H^\circ}{dT} = \Delta C_p$ where $\Delta C_p = C_p^{products} - C_p^{reactants}$. So that:

$$\Delta H^\circ(T) = \Delta H^\circ(298) + \Delta C_p(T - 298)$$

Since in general ΔC_p is not large the approximation $\Delta H^\circ(T) \approx \Delta H^\circ(298)$ is a good one unless T is very different from 298 K.

So given that $\Delta H_f^\circ(298)$ is -242 kJ $K^{-1}mol^{-1}$ we find $\Delta H_f^\circ(600)$ is -245 kJ mol^{-1}. The latter value is not too different from the former. The temperature insensitivity is seen to arise from the relatively small values of the heat capacities, and especially their differences.

Question 20

Enthalpy changes are defined as heat changes at *constant pressure*. However most enthalpies of combustion are measured using an adiabatic bomb calorimeter (as in Questions 5 and 6), in which the reaction is conducted at *constant volume*. Is this difference important?

Answer

If a chemical reaction is conducted at constant volume the heat change is known as the internal energy change, ΔU. In general when reactions are conducted under such conditions the pressure will change during the reaction. As an example consider the combustion examined in Question 6 carried out with stoichiometric amounts of C_3H_6 and O_2:

$$C_3H_6(g) + \frac{9}{2}O_2(g) \rightarrow 3H_2O(\ell) + 3CO_2(g)$$

It can be seen that as there is a reduction in the number of moles of gas present (from 5.5 to 3) the pressure will decrease. Neglecting the vapour pressure of the water formed and treating all gases as ideal the final pressure will be approximately $\frac{3}{5.5}$ times that present initially. This combustion is illustrated in Fig. 3.10 together with the constant pressure case where there is a decrease in volume due again to the reduction in number of moles of gas.

Let us introduce a "thought" experiment in which after the constant pressure experiment is conducted, with a heat change ΔH, we then pull the piston in Fig 3.10(b) against the pressure of the atmosphere until the volume of the products and their pressure, P, is the same as after the reaction at

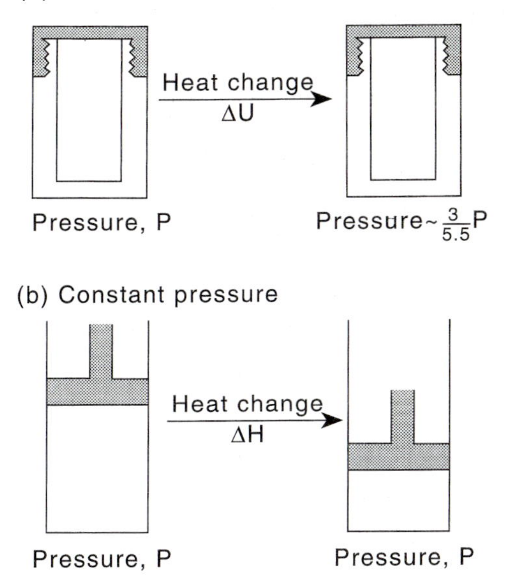

Fig. 3.10 Heat changes under conditions of (a) constant volume and (b) constant pressure.

constant volume. This requires work to be done. The latter quantity is given by (Fig. 3.11):

$$\text{work} = \text{force} \times \text{distance moved}$$

where the force is that exerted on the piston by the atmosphere. This is,

$$\text{force} = P \times A$$

where A is the area of the piston. It follows that

$$\begin{aligned} \text{Work} &= P \times A \times \text{distance moved} \\ &= P\Delta V \\ &= P(V_{\text{reactants}} - V_{\text{products}}). \end{aligned}$$

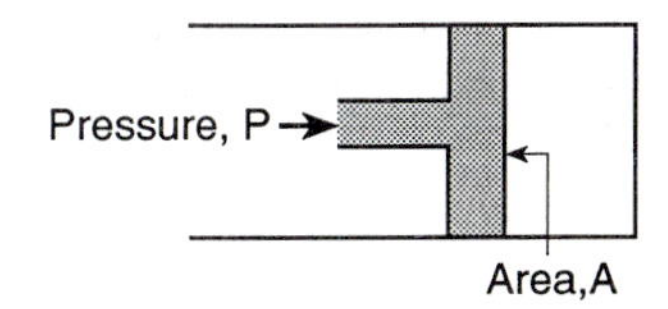

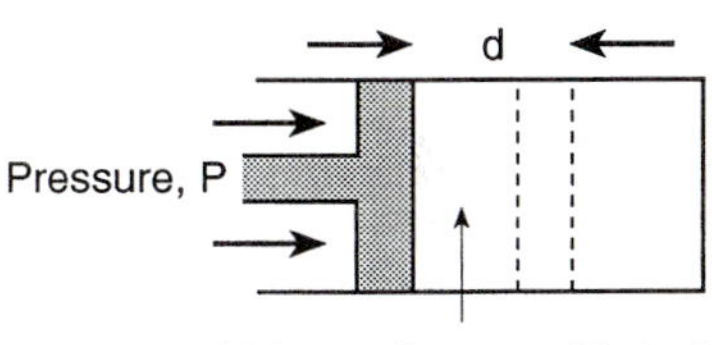

Fig. 3.11 Piston expansion showing a change in volume, ΔV, when reaction occurs constant pressure.

Equating the energy change in the direct constant volume experiment with those in the pair of sequential constant pressure and piston expansion experiments,

$$\Delta U = \Delta H + P(V_{\text{reactants}} - V_{\text{products}})$$

$$\text{or} \quad \Delta H = \Delta U + P(V_{\text{products}} - V_{\text{reactants}}).$$

However as the volume of the products and reactants is overwhelmingly controlled by the amount (number of moles) of gases present,

$$\Delta H = \Delta U + RT\Delta n$$

where we have assumed the gases are ideal and Δn is the change in number of moles of gaseous molecules during the reaction.

Returning to the combustion of cyclopropane we see that the data in Question 6 reveals that

$$\Delta U = -2088 \text{ kJ mol}^{-1}$$

$$\text{In this case} \qquad \Delta n = -2.5 \text{ moles, so that}$$

$$\Delta H = -2088 - 2.5RT$$

$$= -2094 \text{ kJ mol}^{-1}$$

Comparison shows that the difference between ΔU and ΔH is reassuringly small. Finally we note that in the combustion of glucose (Question 5).

$$C_6H_{12}O_6(s) + 6O_2(g) \rightarrow CO_2(g) + H_2O(\ell)$$

the amount of gas is unchanged so $\Delta n = 0$ and $\Delta H = \Delta U$.

Question 21 ΔG° and the Second Law of Thermodynamics

The change in Gibbs free energy, ΔG°, accompanying a chemical reaction is defined by

$$\Delta G^\circ = \Delta H^\circ - T\Delta S^\circ$$

and predicts the direction of change. Specifically if $\Delta G^\circ < 0$ then the reaction proceeds spontaneously. Alternatively prediction may be made on the basis of the Second Law of Thermodynamics which states that 'the entropy of the universe increases in any spontaneous change'. Can these two approaches to the prediction of chemical change be reconciled?

Answer

The Second Law of Thermodynamics tells us that a reaction can only proceed if there is an increase in the entropy of the universe ($S_{universe}$). Mathematically

$$\Delta S_{universe} > 0 \tag{3.2}$$

where ΔS means a change in S. If we consider a reaction vessel and its surroundings, shown in Fig. 3.12,

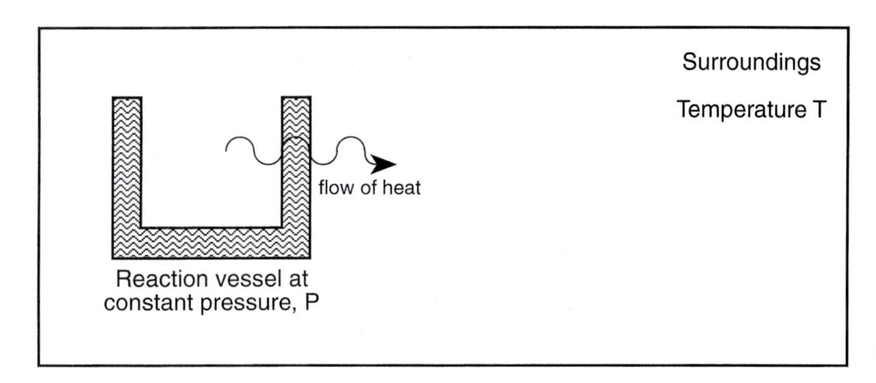

Fig. 3.12 Heat exchanges in the universe between a reaction vessel ("system") and the rest of the universe (the "surroundings").

$$\Delta S_{reaction} + \Delta S_{surroundings} > 0 \tag{3.3}$$

This inequality tells us that the sum of the changes in entropy of the system and the surroundings must be greater than zero.

In the alternative formulation,

$$\Delta G^{\circ}_{reaction} = \Delta H^{\circ}_{reaction} - T\Delta S^{\circ}_{reaction} < 0 \tag{3.4}$$

where the subscript emphasises that we are only considering changes in the chemical reaction of interest. Rewriting equation (3.4),

$$-\Delta G^{\circ}_{reaction} = -\Delta H^{\circ}_{reaction} + T\Delta S_{reaction} > 0$$

or

$$-\frac{\Delta G^{\circ}_{reaction}}{T} = -\frac{\Delta H^{\circ}_{reaction}}{T} + \Delta S_{reaction} > 0 \tag{3.5}$$

Comparison of equations (3.3) and (3.5) shows that the universal entropy and the reaction free energy approaches can be reconciled if

$$\Delta S_{surroundings} = -\frac{\Delta H^{\circ}_{reaction}}{T}.$$

This result will be utilised in the next question.

Question 22: Applying the Second Law to understanding chemical reactions

Atmospheric sulphur dioxide comes from both industrial and natural sources. Natural sources include gases from volcanoes which vent hydrogen sulphide.

When hot volcanic gases mix with air, the hydrogen sulphide can react with oxygen:

$$2H_2S(g) + 3O_2(g) \rightarrow 2H_2O(g) + 2SO_2(g)$$

(a) The sulphur dioxide produced may react with further H_2S emerging from the volcanic vent. This reaction produces crystalline deposits of sulphur. Write a balanced equation for this reaction.

(b) (i) Use the data below to calculate the standard enthalpy change at 298 K for the reaction:

$$2H_2S(g) + 3O_2(g) \rightarrow 2H_2O(\ell) + 2SO_2(g).$$

$$\Delta H^{\circ}_f(H_2S(g)) = -21 \text{ kJ mol}^{-1}$$

$$\Delta H^{\circ}_f(H_2O(\ell)) = -286 \text{ kJ mol}^{-1}$$

$$\Delta H^{\circ}_f(SO_2(g)) = -297 \text{ kJ mol}^{-1}$$

(ii) In the light of Question 21, what is the entropy change in the surroundings at 298 K?

(iii) Use the absolute entropies (at 298 K) below to calculate the entropy change for the reaction.

$$S^{\circ}(H_2S(g)) = +206 \text{ J K}^{-1} \text{ mol}^{-1}$$

$$S^{\circ}(O_2(g)) = +103 \text{ J K}^{-1} \text{ mol}^{-1}$$

$$S^{\circ}(SO_2(g)) = +248 \text{ J K}^{-1} \text{ mol}^{-1}$$

$$S^{\circ}(H_2O(\ell)) = +70 \text{ J K}^{-1} \text{ mol}^{-1}$$

Comment on the sign of the quantity deduced in terms of the Second Law.

Answer

(a) The balanced equation is

$$SO_2(g) + 2H_2S(g) \rightarrow 3S(g) + 2H_2O(\ell)$$

(b) (i) The sought enthalpy change ΔH° may be found either by constructing a Hess cycle or algebraically as follows using the equation shown below.

$$\Delta H^\circ_{reaction} = \sum_{products} \Delta H^\circ_f - \sum_{reactants} \Delta H^\circ_f$$

For the reaction of interest

$$2H_2S(g) + 3O_2(g) \rightarrow 2H_2O(\ell) + 2SO_2(g)$$

$$\begin{aligned}\Delta H^\circ &= 2 \times \Delta H^\circ_f(H_2O(\ell) + 2 \times \Delta H^\circ_f(SO_2(g)) - 2 \times \Delta H^\circ_f(H_2S(g)) \\ &- 3 \times \Delta H^\circ_f(O_2(g)) \\ &= -2 \times 286 - 2 \times 297 - 2 \times (-21) - 3 \times 0 \\ &= -1124 \text{ kJ mol}^{-1}.\end{aligned}$$

(ii) Using the result from Question 21,

$$\Delta S_{surroundings} = -\frac{\Delta H^\circ_{reaction}}{T}$$

so that the entropy change in the surroundings is given by

$$(1124 \times 10^3/298) = 3772 \text{ J K}^{-1} \text{ mol}^{-1}.$$

This is a large and positive value, corresponding to the disorder induced in the surroundings by the release of heat from the reaction.

(iii) The entropy change for the reaction is given by

$$\begin{aligned}\Delta S^\circ &= 2 \times S^\circ(H_2O(\ell) + 2 \times S^\circ(SO_2(g)) - 2 \times S^\circ(H_2S(g)) \\ &- 3 \times S^\circ(O_2(g)) \\ &= 2 \times 70 + 2 \times 248 - 2 \times 206 - 3 \times 103 \\ &= -85 \text{ J K}^{-1} \text{ mol}^{-1}.\end{aligned}$$

The Second Law of Thermodynamics dictates that the entropy of the universe must increase in any feasible chemical reaction:

$$\Delta S_{universe} = \Delta S_{reaction} + \Delta S_{surroundings} > 0$$

In the present reaction the decrease in entropy of the reaction associated with the conversion of five moles of gas into just two moles of gas - is offset by the entropy change of the surroundings. The former quantity is -85 J K^{-1} mol^{-1} where as the latter is 3772 J K^{-1} mol^{-1}.

Question 23: Ellingham diagrams

An Ellingham diagram shows how the Gibbs free energy changes for a particular reaction varies with temperature, T. Fig. 3.13 shows an Ellingham diagram for the following reactions.

(i) $2C(s) + O_2(g) \rightarrow 2CO(g)$ $\quad \Delta G = (-223 - 0.18T)$ kJ mol^{-1}

(ii) $2Fe(s) + O_2(g) \rightarrow 2FeO(s)$ $\quad \Delta G = (-525 + 0.13T)$ kJ mol^{-1}

(iii) $\frac{4}{3}Al(s) + O_2(g) \rightarrow \frac{2}{3}Al_2O_3(s)$ $\quad \Delta G = (-1116 + 0.21T)$ kJ mol^{-1}

(a) Physically, why does ΔG for the formation of CO from C and O_2 become more negative as the temperature increases?
(b) What is the minimum temperature at which reduction of FeO by CO becomes thermodynamically possible?
(c) Aluminium oxide and carbon are both high melting materials. Why is aluminium oxide used in preference to carbon to line vessels that are used to contain molten steel saturated with oxygen? The melting point of steel is around 1800 K.

Answer

(a) The reaction

$$2C(s) + O_2(g) \rightarrow 2CO(g)$$

shows there is an increase in the number of moles of gas so that $\Delta S^\circ > 0$ for this process. With reference to the equation

$$\Delta G^\circ = \Delta H^\circ - T\Delta S^\circ$$

it can therefore be seen that ΔG° becomes more negative as T increases under this inequality.

(b) The minimum temperature at which CO reduces FeO corresponds to the cross over part of the line (i) and (ii) in Fig. 3.13. The temperature at

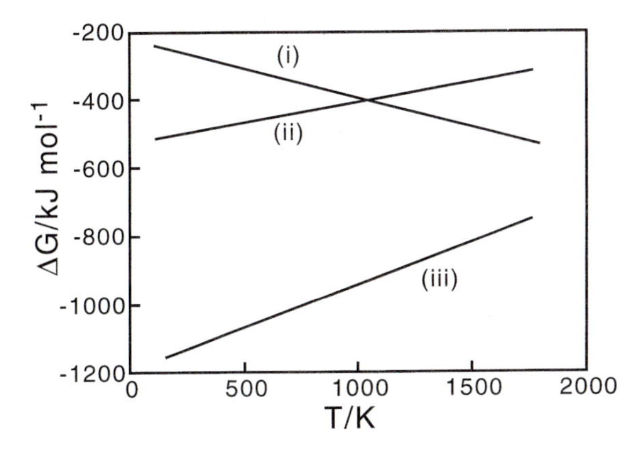

Fig. 3.13 An Ellingham diagram. The individual lines show the dependence of the Gibbs free energy change of a particular reaction with temperature.

which this occurs may be determined accurately by equating the expressions for ΔG° of the two lines:

$$-223 - 0.18\,T = -525 + 0.13\,T$$

so that

$$0.31\,T = 302$$

or

$$T = 974\,K.$$

(c) Inspection of the Ellingham diagram shows that at 1800 K oxygen can react with carbon to form CO since the line (ii) reveals the process to have a negative value of ΔG°. Aluminium, line (iii), will react with oxygen to form Al_2O_3 but this latter material is resistant to further oxidation.

Question 24: Lattice energies and the Born-Landé Equation

Sodium chloride and calcium oxide have the same crystal structure (see Fig. 3.5), with cation-anion distance of 0.235 and 0.239 nm respectively. Using the data below, calculate the lattice energy for both sodium chloride and calcium oxide. Comment on the relative values of the two lattice energies in relation to the charge of each cation, and the cation-anion distance.

Enthalpy of sublimation of sodium, ΔH_{sub}	=	109 kJ mol^{-1}
First ionisation enthalpy of sodium, ΔH_{ion}	=	494 kJ mol^{-1}
First electron affinity of chlorine, ΔH_{EA}	=	−370 kJ mol^{-1}
Enthalpy of dissociation of chlorine, ΔH_{diss}	=	242 kJ mol^{-1}
Standard enthalpy of formation of sodium chloride, ΔH_f°	=	−411 kJ mol^{-1}
Enthalpy of sublimation of calcium, ΔH_{sub}	=	193 kJ mol^{-1}
First ionisation enthalpy of calcium, ΔH_{ion}	=	590 kJ mol^{-1}
Second ionisation enthalpy of calcium, ΔH_{ion}	=	1150 kJ mol^{-1}
First electron affinity of oxygen, ΔH_{EA}	=	−148 kJ mol^{-1}
Second electron affinity of oxygen, ΔH_{EA}	=	850 kJ mol^{-1}
Enthalpy of dissociation of oxygen, ΔH_{diss}	=	496 kJ mol^{-1}
Standard enthalpy of formation of calcium oxide, ΔH_f°	=	−635 kJ mol^{-1}

Answer

The two lattice energies (L.E.) can be calculated from the data given using a Born-Haber cycle. Fig. 3.14 shows this calculation for calcium oxide from which it is clear that

$$(\text{Lattice energy})_{CaO} = 3518\ \text{kJ mol}^{-1}$$

For the case of NaCl a cycle analogous to Fig. 3.2 can be used to deduce that

$$(\text{Lattice energy})_{NaCl} = 787\ \text{kJ mol}^{-1}.$$

The ratio of the two values is 4.5.

To understand the two relative values, consider the sodium chloride structure as shown in Fig. 3.5. We suppose that the charges on the cations and

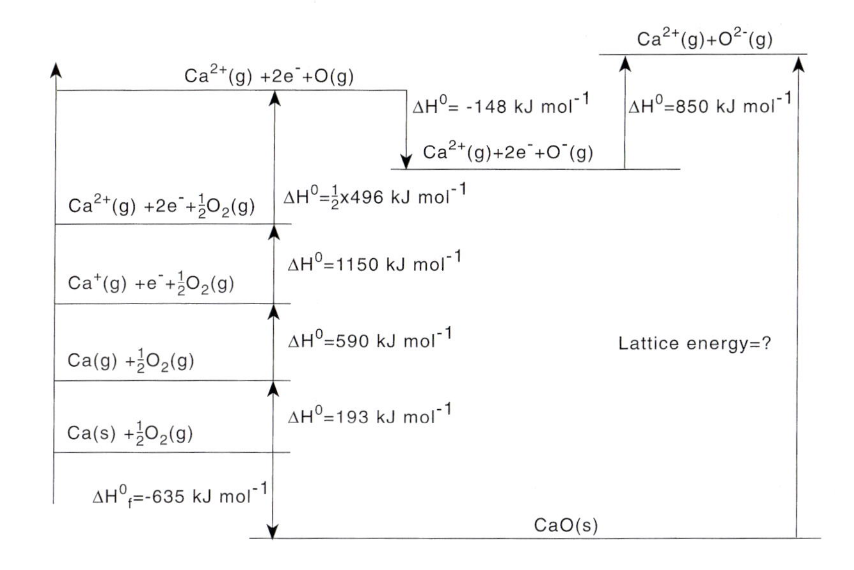

Fig. 3.14 The Born-Haber cycle of CaO.

anions are z_+ and z_- respectively, and treat the ions as hard spheres. The attractive coulombic interaction energy between a pair of such ions separated by a distance r is given by

$$\text{pairwise interaction energy} = -\frac{z_+ z_- e^2}{4\pi\varepsilon_0 r}$$

where ε_0 is the vacuum permittivity and e is the electronic charge. In the crystal, any one ion interacts with *many* others. Specifically for the NaCl lattice (Fig. 3.5), the central ion attracts first six neighbouring oppositely charged ions, second repels twelve like charged ions at the mid points of the

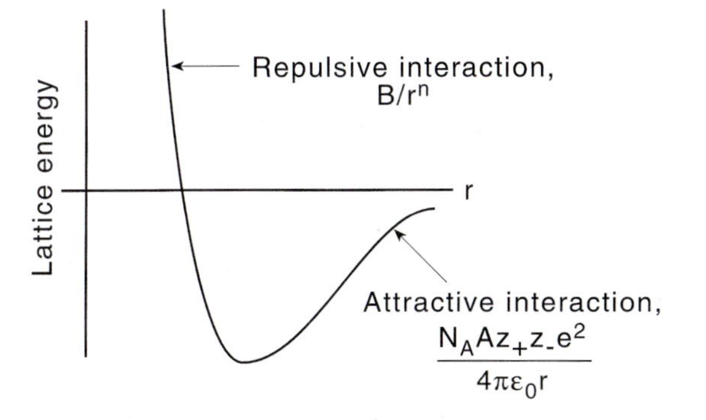

Fig. 3.15 Inter-ionic potential energy plots showing (a) the repulsive and (b) the attractive interactions in an ionic lattice.

edges of the cell shown, third attracts eight oppositely charged ions at the corners, fourth repels the six like ions at the centres of the six adjacent unit cells, fifth..... This sequence forms an infinite series. For a single ion, the interaction energy is,

$$\text{interaction energy} = -\frac{z_+z_-e^2}{4\pi\varepsilon_0 r}\left[6 - \frac{12}{\sqrt{2}} + \frac{8}{\sqrt{3}} - \frac{6}{2} + \ldots\right].$$

For one mole of ions, the net attractive coulombic energy is,

$$\text{net attractive energy} = -\frac{N_A z_+ z_- e^2 A}{4\pi\varepsilon_0 r}$$

where A is the Madelung Constant,

$$A = 6 - \frac{12}{\sqrt{2}} + \frac{8}{\sqrt{3}} - \frac{6}{2} + \ldots = 1.748$$

Madelung constants can be similarly deduced for other unit cell geometries, for example fluorite, CaF_2, 2.519; caesium chloride, CsCl, 1.763; rutile, TiO_2, 2.408.

Of course if these forces discussed above were the only ones operating, the ion-ion separation, r would collapse to a tiny value so as to lower the energy. In practice, the coulombic attraction is opposed by a *repulsive* force which increases very rapidly with decreasing ion-ion separation, *once the ions are in contact*. Next we proceed by assuming the repulsive force varies in the manner $\frac{B}{r^n}$, where n and B are constants, and n is large. It follows that the lattice energy is given by

$$\text{lattice energy} = -\frac{N_A z_+ z_- e^2 A}{4\pi\varepsilon_0 r} + \frac{B}{r^n}$$

Different models for the lattice energy can be developed using different forms of the repulsive interaction. For example, the Born-Mayer expression can be generated using an exponential dependence of the repulsions on r. Kapustinskii noticed that the ratio vA/r is approximately constant for a variety of different unit cells, where v is the number of ions in each unit cell. This enabled the derivation of a general equation to estimate the lattice energy of any simple crystal, without having to know its crystal structure,

$$\text{lattice energy} = -\frac{1.38 \times 10^{-9} z_+ z_- v}{r_+ + r_-}$$

where r_+ and r_- are the ionic radii of the cation and anion respectively.

The variation with r is sketched in Fig. 3.15. The minimum can be shown (by differentiation) to be at a value

$$\text{lattice energy} = \frac{N_A A z_+ z_- e^2}{4\pi\varepsilon_0 r}\left[1 - \frac{1}{n}\right]$$

Returning to the comparison of NaCl and CaO it can be seen from the above equation that, as the radius is similar for both metals, the ratio

$$\frac{(\text{L.E.})_{\text{CaO}}}{(\text{L.E.})_{\text{NaCl}}} \approx 4$$

given that the ion–ion separation is about equal in the two materials. This is close to the ratio deduced above. The deviation from four observed experimentally arises partly from the variation of the exponent n from one material to another but also because the bonding may not be entirely ionic, some degree of covalent interaction may be present, particularly in CaO.

4 Chemical kinetics

4.1 Aims

In this chapter, the following basic topics are addressed in section 4.2.

- Rates of reaction and their concentration dependence.
- The order of a chemical reaction.
- The relationship of reaction mechanisms to kinetic rate laws.
- The temperature dependence of chemical reaction rates.
- Examples of the kinetics and mechanism of solution and gas phase reactions.

Section 4.3 visits the following more advanced material.

- Integrated rate laws.
- Kinetics of gas phase reactions as revealed by pressure measurements.
- The steady state approximation.
- Reaction via tunnelling, not activation.
- The use of the Arrhenius equation to predict the vapour pressures of gases; the Clausius-Clapeyron equation.

4.2 Problems on core topics

Question 1

The reaction between hydroxide ions, OH^-, and phosphinate ions, $H_2PO_2^-$,

$$H_2PO_2^-(aq) + OH^-(aq) \rightarrow H_2(g) + HPO_3^{2-}(aq)$$

has been studied by following the initial rate of formation of the gaseous product. The results are tabulated below.

Initial concentration of $H_2PO_2^-(aq)$ / mol dm^{-3}	Initial concentration of $OH^-(aq)$ / mol dm^{-3}	Initial rate of $H_2(g)$ evolution / 10^{-3} dm^3 min^{-1}
0.6	1.0	2.3
0.6	2.0	9.5
0.6	3.0	21.6
0.1	6.0	14.4
0.2	6.0	28.7
0.3	6.0	43.2

The structures of $H_2PO_2^-$ and HPO_3^{2-} are shown below. Both compounds have a distorted tetrahedral distribution of atoms around the P atom.

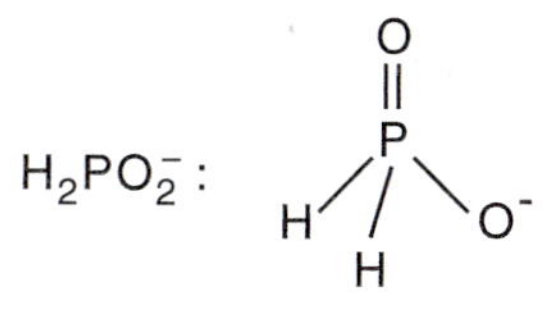

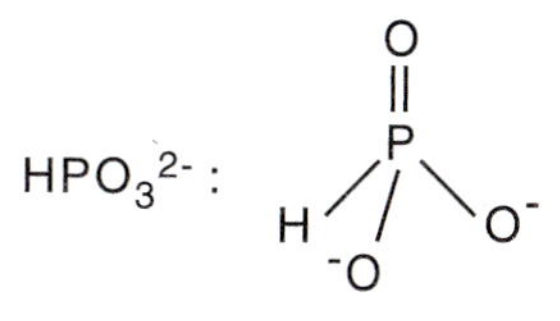

(a) What is the order of the reaction with respect to $OH^-(aq)$ ions?
(b) What is the order of the reaction with respect to $H_2PO_2^-(aq)$ ions?
(c) Write the rate expression for this reaction.
(d) What are the units of the rate constant?

Answer

(a) In comparing the first three sets of data, note that the initial concentration of $H_2PO_2^-$ remains constant whilst the relative hydroxide concentration changes from 1 to 2 to 3. The rate of hydrogen evolution changes from 2.3 to 9.5 to 21.6 $cm^3\ min^{-1}$. These latter values are in the ratio 1:4.1:9.4. This suggests that the rate of reaction is,

$$R \propto [OH^-]^2.$$

(b) To find the order with respect to $H_2PO_2^-$, the last three sets of data are helpful since the hydroxide concentration is maintained constant. As the concentration of $H_2PO_2^-$ increases from 0.1 to 0.2 to 0.3, the rate of hydrogen evolution alters from 14.4 to 28.7 to 43.2. These values are almost exactly in the ratio 1 to 2 to 3. This suggests that the order of reaction is one:

$$R \propto [H_2PO_2^-].$$

A not implausible mechanism coinciding with the observed rate law is:

The concentrations used in this study are very high. Normally significantly lower values are used if at all possible.

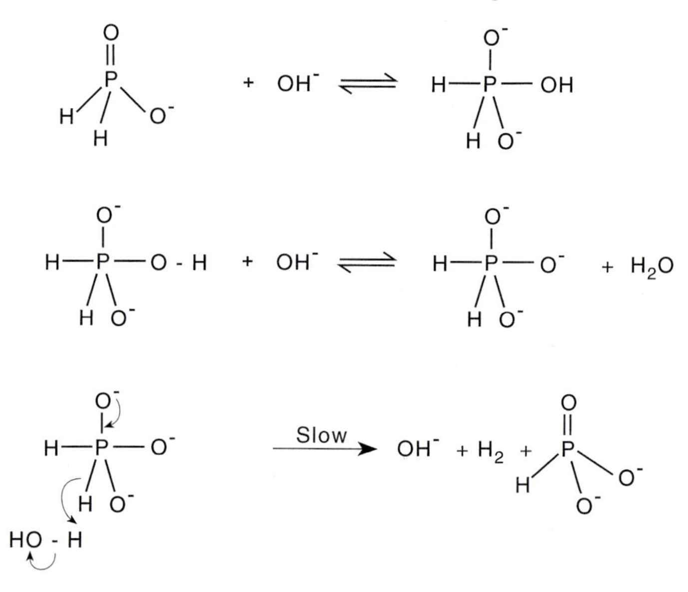

(c) The full rate expression is:

$$R = k[H_2PO_2^-][OH^-]^2 \tag{4.1}$$

where k is the rate constant for the reaction.

(d) The units of the rate constant can be found by noting the units of the other quantities in the rate equation (4.1):

$R/dm^3\ min^{-1}$
$[OH^-]/mol\ dm^{-3}$
$[H_2PO_2^-]/mol\ dm^{-3}$.

It follows that k has units of $mol^{-3}\ dm^{12}\ min^{-1}$.

Question 2

The rate of the reaction

$$2NO(g) + O_2(g) \rightarrow 2NO_2(g)$$

was studied near 70 K by varying the initial partial pressures, p, of the reactants and measuring the initial rates. The results are tabulated below.

Initial pressure of $NO(g)$ / N m^{-2}	Initial pressure of $O_2(g)$ / N m^{-2}	$10^8 \times$ Initial rate of formation of $NO_2(g)$ / N m^{-2} s^{-1}
1630	1630	6.13
3260	1630	24.5
1630	3260	12.2

This reaction can be followed using a colorimeter: NO is colourless, whilst NO_2 is a dark brown gas. Recent research has evolved "electronic noses" which can characterise and quantify materials though their odour; NO_2 is pungent and would be amenable to detection in this manner. For a discussion of electronic noses and their application in the food industry see P. N. Bartlett *et al*, *Food Technology*, 51 (1997) 44.

(a) What is the rate equation for this reaction?
(b) Calculate the rate constant.
(c) Draw a reaction profile.

Answer

(a) Inspection of the data shows that the rate,

$$R \propto p(NO)^2 \quad (4.2)$$

and

$$R \propto p(O_2) \quad (4.3)$$

so that the overall rate law is

$$R = k\, p(NO)^2 p(O_2) \quad (4.4)$$

where k / N^{-2} m^4 s^{-1} is the rate constant.

(b) By substituting the three sets of data into equation (4.4) and averaging the three resulting values of k, it is found that

$$k = 1.4 \times 10^{-17}\ N^{-2}\ m^4\ s^{-1}.$$

(c) The kinetics imply that the formation of the transition state for the reaction involves the prior reaction of two molecules of NO and one of O_2. Since the simultaneous reaction of all three molecules is highly improbable, sequential reaction of the molecules is presumed. As NO is a molecule with an unpaired electron in its ground molecular electronic state, a plausible reaction scheme involves the formation of the dimer, $(NO)_2$, in an initial step.

$$2NO(g) \rightleftharpoons (NO)_2(g)$$

$$(NO)_2(g) + O_2(g) \xrightarrow{\text{slow}} 2NO_2(g).$$

The reaction profile is shown in Fig. 4.1. Notice that the first step is drawn as a pre-equilibrium, so that $(NO)_2$ is formed from NO and dissociates into the same, many more times than reaction with oxygen occurs.

Question 3

It has been found that for the reaction,

$$2NO(g) + 2H_2(g) \rightarrow N_2(g) + 2H_2O(g)$$

the rate doubles as the pressure of hydrogen is doubled at a constant pressure of NO(g). However, the rate goes up by a factor of four if the pressure of nitrogen oxide is doubled whilst maintaining a constant pressure of hydrogen.

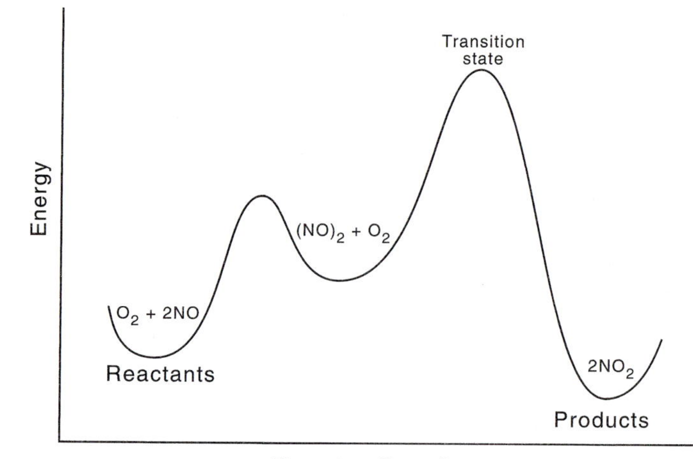

Fig. 4.1 The reaction profile for the reaction

$$2NO(g) \rightleftharpoons (NO)_2(g)$$
$$(NO)_2(g) + O_2(g) \xrightarrow{\text{slow}} 2NO_2(g).$$

(a) What is the rate equation for the reaction? Give the units of all the terms in the expression.

(b) What is the order of the reaction. What can be inferred from comparing the order with respect to hydrogen and the stoichiometric coefficient for hydrogen in the chemical equation?

(c) Draw a reaction profile, labelling an example of an intermediate and a transition state.

Answers

(a) The rate is given by

$$R = k\,p(NO)^2 p(H_2)$$

where the units are R / N m^{-2} s^{-1}, p / N m^{-2} and k / N^{-2} m^4 s^{-1}.

(b) The overall order of the reaction is three. The order with respect to NO is two and that with respect to H_2 is one.

The order with respect to hydrogen shows that one molecule of this species participates in the transition state. The second molecule of hydrogen, known to react from the reaction stoichiometry, must react after the transition state.

(c) This reaction occurs in three steps. First a rapid pre-equilibrium occurs forming a $(NO)_2$ dimer. The latter species is an *intermediate* which reacts slowly with hydrogen, via a *transition state* to form nitrogen gas and another *intermediate species*, H_2O_2. Subsequently, the latter reacts with more hydrogen to form the products. This sequence of events is described below.

$$2NO(g) \rightleftharpoons (NO)_2(g)$$
$$(NO)_2(g) + H_2(g) \xrightarrow{\text{slow}} N_2(g) + H_2O_2(g)$$
$$H_2O_2(g) + H_2(g) \xrightarrow{\text{fast}} 2H_2O(g).$$

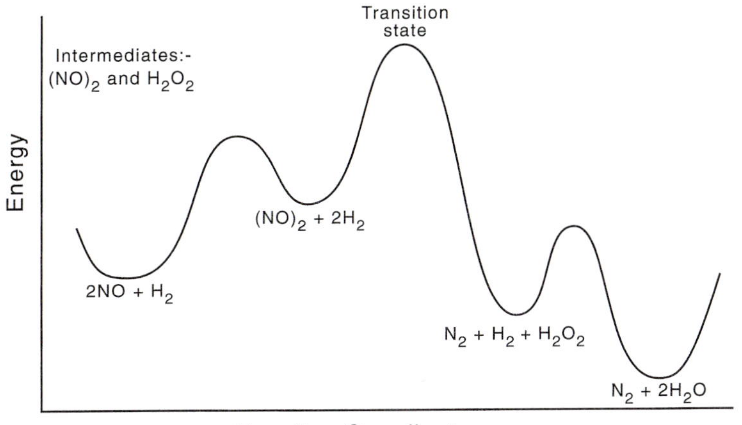

Fig. 4.2 The reaction profile for the reaction

$$2NO(g) \rightleftharpoons (NO)_2(g)$$
$$(NO)_2(g) + H_2(g) \xrightarrow{\text{slow}} N_2(g) + H_2O_2(g)$$
$$H_2O_2(g) + H_2(g) \xrightarrow{\text{fast}} 2H_2O(g).$$

Note that the first step is a pre-equilibrium so that $(NO)_2$ is formed from NO and decays back into NO many times before the reaction with hydrogen occurs.

The reaction profile is shown in Fig. 4.2.

Question 4

The stoichiometric equation for the iodination of propanone in an aqueous solution of sulphuric acid is:

$$CH_3COCH_3(aq) + I_2(aq) \rightarrow CH_2ICOCH_3(aq) + H^+(aq) + I^-(aq).$$

The kinetics of this reaction were studied by withdrawing samples from the reaction mixture, quenching them, and analysing the remaining iodine. The following results were obtained.

Experiment	Volume of propanone	Volume of 1M H_2SO_4 / cm^3	Volume of H_2O / cm^3	10^4 reaction rate / mol dm^{-3} min^{-1}
1	20	10	150	1.6
2	15	10	155	1.2
3	10	10	160	0.8
4	20	15	145	2.4
5	20	5	155	0.8

(a) Deduce the order of the reaction with respect to propanone and sulphuric acid.

(b) The rate of reaction was found to be zero order with respect to iodine. Give the rate equation and suggest a possible mechanism for the reaction.

(c) Draw a reaction profile.

Answer

(a) Inspection of the data shows that the rate of reaction is first order with respect to propanone and is first order with respect to sulphuric acid. The overall reaction order is therefore two.

(b) If the order with respect to iodine is zero, and the order with respect to propanone and to acid is one, then the rate of reaction is:

$$R \propto [CH_3COCH_3]^1[H^+]^1[I_2]^0.$$

Any mechanism which is consistent with this observed rate law must first involve one molecule each of propanone and H^+ in the transition state and, second, implicate iodine in the mechanism only after the formation of the transition state. The accepted mechanism involves the formation of the enol form of propanone.

If the sulphuric acid is fully dissociated then $[H^+]$ is twice the concentration of H_2SO_4.

In this reaction H^+ is a catalyst. Catalysts alter the rate at which a chemical reaction occurs and can be recovered chemically unchanged at the end of the reaction. Usually catalysts accelerate the rate of reaction (positive catalysis) but there also examples of negative catalysis whereby the rate of the reaction is decelerated. There are two forms of catalyst; homogeneous, where the catalyst is in the same phase as the reactant (for example enzymes) and heterogeneous, where the catalyst is in a different phase (for example palladium catalysis in hydrogenation of alkenes).

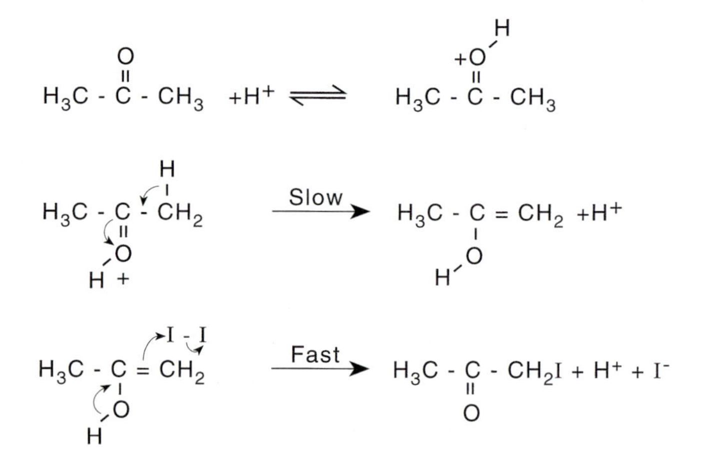

Notice that enol formation must be irreversible; if both the first two steps were pre-equilibria, then there would be no net consumption of protons (H^+) before the transition state, and the first order observation of $[H^+]$ would be absent.

(c) The reaction profile is shown in Fig. 4.3.

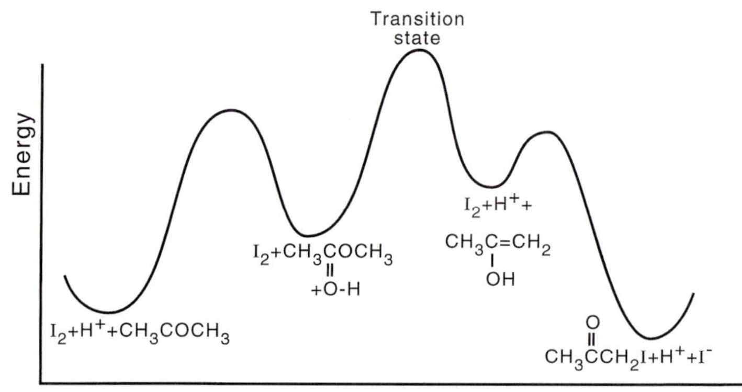

Fig. 4.3 The reaction profile for the iodination of propanone.

$$CH_3COCH_3(aq) + I_2(aq) \rightarrow CH_2ICOCH_3(aq) + H^+(aq) + I^-(aq).$$

Question 5

The gas phase reaction between CO and Cl_2 produces phosgene,

$$CO(g) + Cl_2(g) \rightarrow COCl_2(g).$$

The rate equation can be written as:

$$\text{rate} = kp(CO)^x p(Cl_2)^y,$$

where p(CO) and $p(Cl_2)$ are the partial pressures of CO and Cl_2 respectively, and x and y are the respective reaction orders. The following kinetic measurements have been made.

Recall that the partial pressure of a species is directly proportional to its concentration.

Rate / mm Hg s^{-1}	p(CO) / mm Hg	$p(Cl_2)$/mm Hg
0.131	100	120
0.329	250	120
0.590	450	120
0.372	100	240
0.683	100	360

(a) What is the order of the reaction with respect to Cl_2 and to CO?
(b) Write down a rate law for the reaction and suggest a mechanism consistent with this.

Answer

(a) Inspection of the data shows that the reaction has order one with respect to CO and order $\frac{3}{2}$ with respect to Cl_2.
(b) The rate law shows that the rate of reaction,

$$R = k\ p(CO)^1\ p(Cl_2)^{1.5}.$$

A mechanism consistent with this is:

$$Cl_2(g) \rightleftharpoons 2Cl^\bullet(g) \qquad K_1$$

$$Cl^\bullet(g) + CO(g) \rightleftharpoons {}^\bullet COCl(g) \qquad K_2$$

$${}^\bullet COCl(g) + Cl_2(g) \xrightarrow{\text{slow}} COCl_2(g) + Cl^\bullet(g).$$

The first two steps are pre-equilibria so that

$$K_1 = \frac{p(Cl^\bullet)^2}{p(Cl_2)},$$

and

$$K_2 = \frac{p({}^\bullet COCl)}{p(Cl^\bullet)p(CO)}.$$

In the slow stage of the reaction the $Cl^\bullet$ radical is regenerated; this is an example of a *chain reaction.*

It follows that

$$p({}^\bullet COCl_2) = K_2\ p(Cl^\bullet)\ p(CO)$$
$$= K_1^{\frac{1}{2}}\ K_2\ p(CO)\ p(Cl_2)^{\frac{1}{2}}.$$

Thus the rate of the slow step is

$$R \propto p(Cl_2)p({}^\bullet COCl)$$
$$\propto K_1^{\frac{1}{2}} K_2 p(CO) p(Cl_2)^{1.5},$$

consistent with the observed kinetics.

Question 6

Oxidation of methanal in aqueous acidic solution by the VO_2^+ ion obeys the rate law:

$$-\frac{d[VO_2^+]}{dt} = k\,[CH_2O]\,[VO_2^+]^2\,[H^+]^2,$$

where $-\dfrac{d[VO_2^+]}{dt}$ is the rate of disappearance of VO_2^+.

(a) What is the overall reaction order?
(b) What are the units of the rate constant, k?
(c) How can the appearance of so many molecules in the rate equation be accounted for?
(d) Suggest a mechanism for the reaction consistent with the observed rate law.

Answer

(a) The reaction has order one with respect to $[CH_2O]$, order two with respect to $[VO_2^+]$ and order two with respect to $[H^+]$. The overall order is therefore five.
(b) The rate constant has units of $mol^{-4}\ dm^{12}\ s^{-1}$.
(c) The appearance of so many molecules in the rate equation is strongly indicative of one, or more, pre-equilibria in the mechanism occuring before the transition state.
(d) A not implausible mechanism is that shown below. Note that methanal, CH_2O, exists in aqueous solution, exclusively as the hydrated form, $CH_2(OH)_2$ whilst the VO_2^+ ion participates in the equilibrium,

$$VO_2^+(aq) + H^+(aq) + H_2O(\ell) \rightleftharpoons V(OH)_3^{2+}(aq).$$

Note also that the transition state is formed by the reaction of one mole of methanal, two VO_2^+ and two protons.

$HCH(OH)_2 + V(OH)_3^{2+} \rightleftharpoons$ HCHOH with HO: $\rightarrow V(OH)_3^{2+}$ (fast)

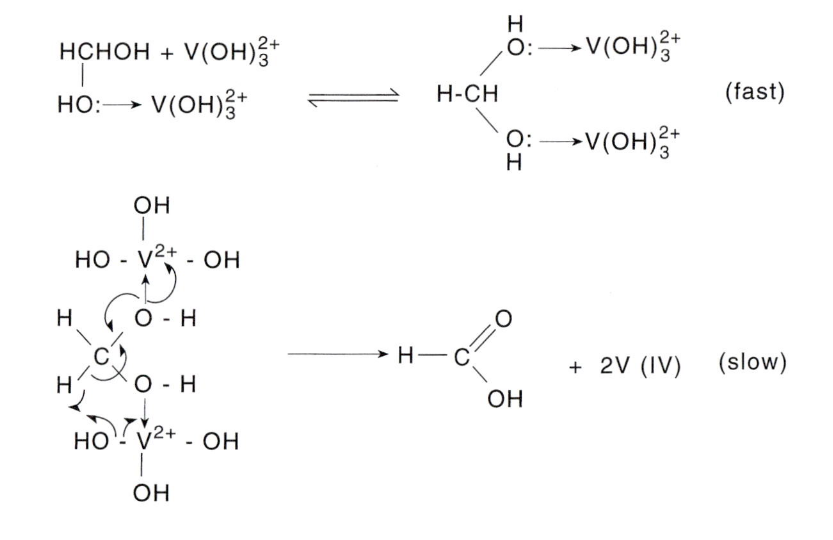

A fishhook implies the transfer of only one electron in the same way as a curly arrow denotes the movement of an electron pair.

Question 7

(a) Explain how and why the rate of a bimolecular reaction in the gas phase depends on the temperature. Justify, for the case of a gas phase reaction, the form of the Arrhenius equation,

$$k = A\exp(-E_a / RT).$$

(b) Given $A = 1 \times 10^{11}\ dm^3\ mol^{-1}\ s^{-1}$ and $T = 298$ K, why do the following second order rate constants differ?

	$k / mol^{-1} dm^3\ s^{-1}$
$2I(g) \rightarrow I_2(g)$	7×10^9
$2NOBr(g) \rightarrow 2NO(g) + Br_2(g)$	7×10^{-1}

Answer

(a) The requirements for a successful bimolecular reaction between molecules A and B in the gas phase are as follows.

- A collision between A and B is essential.
- The collision must be adequately energetic to overcome the activation energy of the reaction.
- The reacting species must have the correct relative orientation when colliding for the reaction to occur.

These facts can be grouped together in the following expression for the bimolecular reaction rate.

$$R = (C_{AB}[A][B])(\exp(-E_a / RT))(p) \qquad (4.5)$$

where the first term $C_{AB}[A][B]$ describes the number of collisions between A and B per unit volume in unit time and C_{AB} is related to the relative velocity of the molecules A and B. The second term describes the fraction of the collisions which have an energy in excess of the activation energy, E_a; the quantity $\exp(-E_a / RT)$ ranges from near unity for a small value of E_a to almost zero for large activation energy values. Finally the third factor, p, the so called steric factor, accounts for the probability that reactant molecules have the correct relative orientation for reaction. By comparison of equation (4.5) with the second order reaction law,

$$R = k_2[A][B]$$

it can be seen that

$$k_2 = A\exp(-E_a / RT)$$

where the pre-exponential factor A involves both relative velocity and steric factor terms.

(b) Applying the Arrhenius equation for the reaction

$$2I(g) \rightarrow I_2(g)$$

we see that

$$7 \times 10^9 = 10^{11}\exp(-E_a / 8.31 \times 298)$$

which implies that the activation energy is 7 kJ mol^{-1}.
Similarly for the reaction

$$2NOBr(g) \rightarrow 2NO(g) + Br_2(g)$$

it can be found that the activation energy is 64 kJ mol^{-1}. The difference between the two energies can be rationalised since in the first reaction there is bond making whereas in the second reaction bonds must break.

Question 8

Methane is an important greenhouse gas. One path for its removal from the lower atmosphere is the reaction:

$$HO^{\bullet}(g) + CH_4(g) \rightarrow H_2O(g) + CH_3^{\bullet}(g).$$

This reaction has second order kinetics, with an activation energy, $E_a = 21$ kJ mol^{-1}. What is the ratio of the rate constant at a point on the Earth's surface, where the temperature is 298 K, to that at a point five miles above the Earth's surface, where the temperature is about 260 K?

"The greenhouse effect" is the phenomenon whereby heat emitted by the Earth is unable to escape into the rest of the solar system, thus keeping the Earth warm. Atmospheric gases such as carbon dioxide, methane and, to a lesser extent, sulphur dioxide and various nitrogen oxides, cause this effect by effectively "trapping" the heat in various bond stretching or bending vibrations within the molecule. This trapped energy is subsequently re-radiated to the Earth.

Answer

The two rate constants can be predicted from the Arrhenius equation,

$$k = A\exp(-E_a / RT)$$

so that

$$\frac{k_{298}}{k_{260}} = \exp\left(-\frac{E_a}{R}\left(\frac{1}{298} - \frac{1}{260}\right)\right)$$
$$= 3.45$$

Note that A is assumed to be independent of temperature. In reality this is an approximation—but a good one—since the temperature dependence of the relative velocity of the reactants is approximately $T^{\frac{1}{2}}$. In the present example the error in treating A as constant can therefore be seen to be in the order of $(298/260)^{\frac{1}{2}}$ corresponding to approximately 5%.

Question 9

The rate constant, k, of the (overall) second order reaction

$$H_2(g) + I_2(g) \rightarrow 2HI(g)$$

varies with temperature as follows

T / K	556	700	781
k / mol^{-1} dm^3 s^{-1}	1.19×10^{-4}	1.72×10^{-1}	3.58

Calculate:

(a) the activation energy of the reaction,
(b) the rate constant at 629 K, and

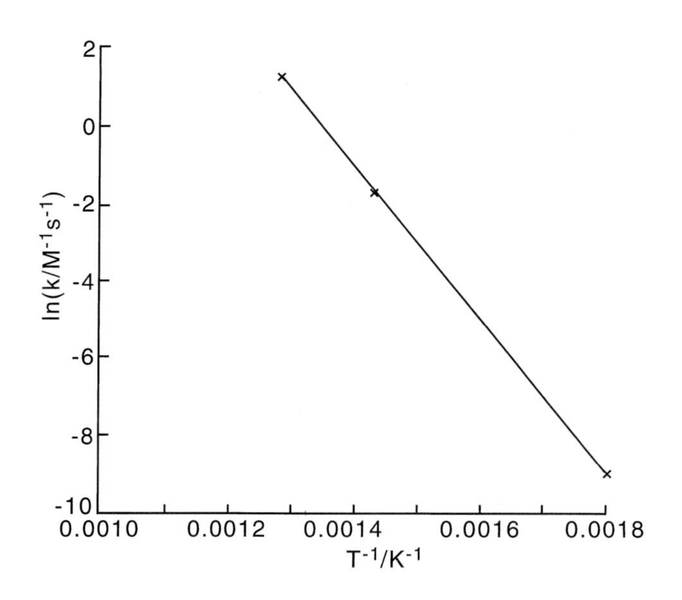

Fig. 4.4 A graph to show ln k against T^{-1} for the data in Question 9.

(c) the rate constant for the reverse reaction at 629 K given that the equilibrium constant is 3.73 at this temperature.

Answers

(a) The activation energy can be found using the Arrhenius equation.

$$k = A \exp(-E_a / RT)$$

so that

$$\ln k = \ln A - E_a / RT$$

Accordingly a plot of ln k against (1 / T) is linear with slope of $(-E_a / R)$. Such a plot for the data in the question is given in Fig 4.4. The slope has a value of

$$-E_a / R = -19913 \text{ K},$$

so that

$$E_a = 165 \text{ kJ mol}^{-1}.$$

(b) The rate constant at 629 K can be found from inspection of Fig. 4.4 by working out the value of (1 / T) corresponding to a temperature of 629 K and finding the associated value of ln k. The rate constant k is found to be 7.29×10^{-3} mol^{-1} dm^3 s^{-1}.

(c) At equilibrium

$$H_2(s) + I_2(g) \underset{k_{-1}}{\overset{k_1}{\rightleftarrows}} 2HI(g),$$

we know that the rate of the formal reaction must balance that of the reverse reaction, so that

$$k_1[H_2][I_2] = k_{-1}[HI]^2.$$

It follows that

$$\frac{k_1}{k_{-1}} = \frac{[HI]^2}{[H_2][I_2]} = K_{eq}$$

where K_{eq} is the equilibrium constant for the reaction. If the latter has a value of 3.73 and if k_1 is known to be 7.29×10^{-3} mol^{-1} dm^3 s^{-1}, then

$$k_{-1} = 1.95 \times 10^{-3} \text{ mol}^{-1} \text{ dm}^3 \text{ s}^{-1}.$$

4.3 Taking it further

Question 10: Integrated rate laws

In a reaction mixture at a fixed temperature, the concentration of a species, [A], varies with time, t, in the following manner.

t / s	0	500	1000	2000	3000	4000
[A]/mol dm^{-3}	1.0	2.1×10^{-6}	1.0×10^{-6}	4.8×10^{-7}	3.3×10^{-7}	2.5×10^{-7}

In section 4.2 measurements of *initial rates* were used to establish rate laws. Usually, however, it is more reliable to monitor the concentrations of the reactants over a period of time, so as to establish the kinetic behaviour.

(a) Construct linear plots to determine whether the reaction is first or second order overall. Calculate the rate constant and give an expression for the reaction rate.

(b) What is the concentration of A after three hours have elapsed since the reaction started?

(c) How does the half-life of A vary with time?

Answer

(a) We start with the case of first order kinetics,

$$A \rightarrow \text{products}$$

where the rate law is,

$$\text{rate} = \frac{d[A]}{dt} = -k_1[A] \qquad (4.6)$$

In the equation, k_1/s^{-1}, is the first order rate constant and $\frac{d[A]}{dt}$ is the first derivative of the concentration with respect to time.

To establish how [A] varies with time, the rate law must be integrated,

$$\int \frac{d[A]}{[A]} = \int -k_1 dt \qquad (4.7)$$

from which it follows that

$$\ln [A] = -k_1 t + c \qquad (4.8)$$

where c is a constant. Now, at the start of the experiment, $t = 0$, the concentration of A is $[A]_0$. Substitution of this "boundary condition" into equation (4.8) shows that

$$\ln [A]_0 = c \tag{4.9}$$

We therefore conclude that

$$\ln [A] - \ln [A]_0 = -k_1 t \tag{4.10}$$

or

$$\ln([A]/[A]_0) = -k_1 t \tag{4.11}$$

or

$$[A] = [A]_0 \exp(-k_1 t) \tag{4.12}$$

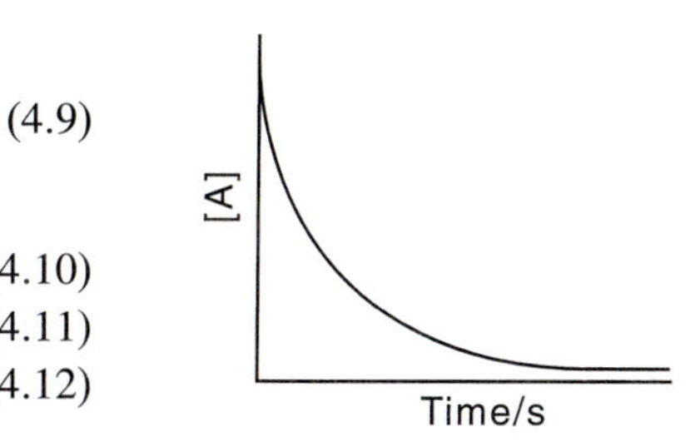

Fig. 4.5 The exponential decay of [A] during a first order process.

Equation (4.12) shows that [A] decreases exponentially for a first order reaction (Fig. 4.5). In practice, when analysing experimental data (compilations of [A] as a function of time) equation (4.8) is much more useful, since it converts the data into a straight line plot if the data is indeed compatible with first order kinetics: equation (4.8) suggests a plot of ln [A] against time will be linear in such cases (Fig. 4.6).

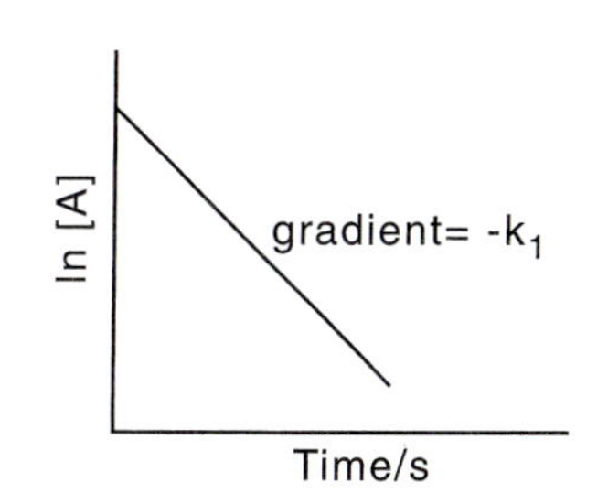

Fig. 4.6 Integrated rate law to test for first order kinetics: a linear plot of ln [A] against time is consistent with such behaviour. The slope of the plot ($-k_1$) allows the deduction of the first order rate constant, k_1.

Fig. 4.7 shows the analysis of the data in the question in terms of first order kinetics. The plot is *not* linear, so it must be concluded that first order kinetics cannot be operating. Accordingly, we next turn to consider second order kinetics.

For second order kinetics, the rate law is

$$\frac{d[A]}{dt} = -k_2[A]^2 \tag{4.13}$$

where k_2/mol^{-1} dm^3 s^{-1} is the second order rate constant. Again the law must be integrated to discover the variation of [A] with time

$$\int \frac{d[A]}{[A]^2} = \int -k_2 dt \tag{4.14}$$

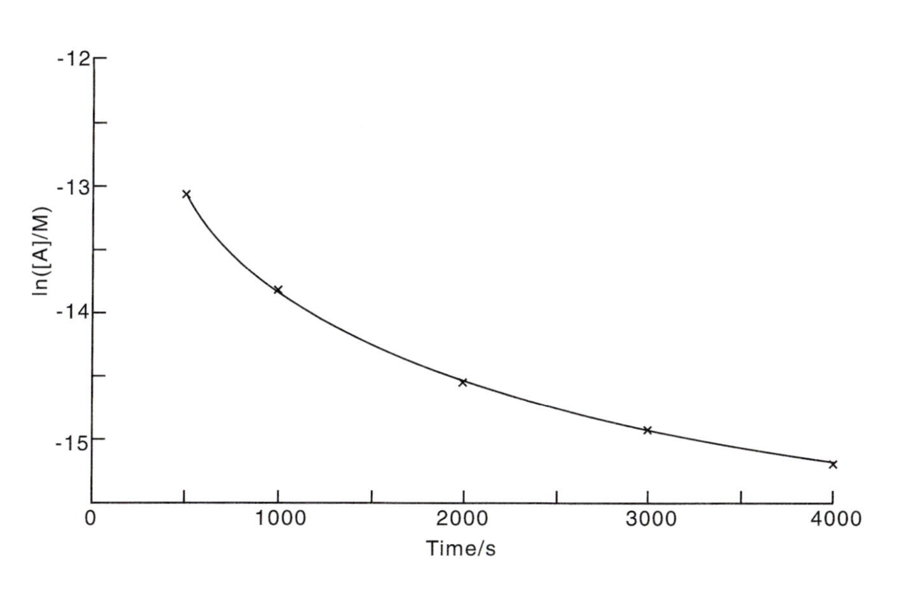

Fig. 4.7 A graph showing analysis of the data in Question 10 if first order kinetics are assumed to apply.

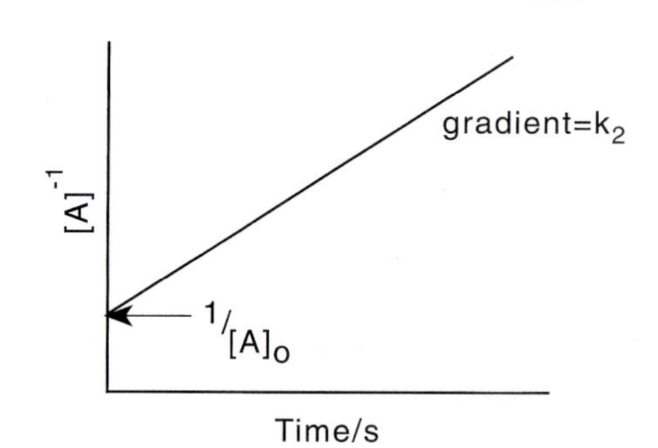

Fig. 4.8 Integrated rate law testing second order kinetics: a linear plot of $[A]^{-1}$ against time suggests such behaviour. The slope of the line is k_2.

Performing the integration gives

$$\frac{1}{[A]} = k_2 t + c \tag{4.15}$$

where c is again a constant. Since the initial concentration of A, at $t = 0$, is $[A]_0$, it can be seen that

$$\frac{1}{[A]_0} = c \tag{4.16}$$

The full integrated rate law is

$$\frac{1}{[A]} - \frac{1}{[A]_0} = k_2 t \tag{4.17}$$

Equation (4.17) shows that to test for second order kinetics, a plot of $[A]^{-1}$ against time should be made. If the plot is linear, the data is not inconsistent with second order kinetics (Fig. 4.8). Fig. 4.9 show the experimental data

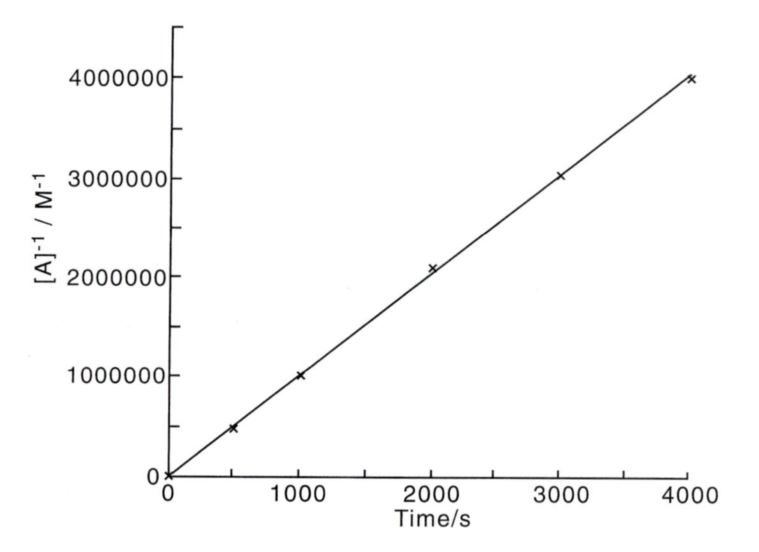

Fig. 4.9 The experimental data in Question 10 gives a linear plot if second order kinetics are assumed.

analysed in terms of a second order kinetics. A good straight line is evident, implying that the reaction is second order. The slope of the line is 10^3 mol^{-1} dm^3 s^{-1}which corresponds to the rate constant.
The rate equation for the reaction is therefore

$$\frac{d[A]}{dt} = -10^3[A]^2.$$

(b) We first convert the time of three hours into seconds:

$$\text{time} = 3 \times 60 \times 60 = 10800\text{s}.$$

Using equation (4.17),

$$\frac{1}{[A]} - \frac{1}{[A]_0} = k_2 t = 10^3 t = \frac{1}{[A]} - \frac{1}{1.0},$$

so that when $t = 10800$ s, $[A] = 9.26 \times 10^{-8}$ mol dm^3.

(c) For a second order reaction at a time corresponding to the half-life, $t = t_{\frac{1}{2}}$, we see that

$$\frac{1}{[A]_0/2} - \frac{1}{[A]_0} = k_2 t_{\frac{1}{2}}$$

so that

$$\frac{1}{k_2[A]_0} = t_{\frac{1}{2}}.$$

It follows that the half-life increases as the solution gets more dilute. Inspection of the data in the Question shows this to be the case.

A quick test to whether a reaction is first order or not, is to examine the *half-life*, $t_{\frac{1}{2}}$, of the reactant. This is the time taken for the concentration of the reactant to decrease by a factor of two. Equation (4.11) shows that $t_{\frac{1}{2}} = \ln 2/k_1$, so that for first order kinetics, $t_{\frac{1}{2}}$ is *independent of concentration*. This is not true for other reaction orders.

Question 11: Integrated rate laws: first order kinetics

The kinetics of the decomposition of azomethane were studied by determining the time dependence of its partial pressure, p, at 550 K:

$$CH_3N_2CH_3(g) \rightarrow CH_3CH_3(g) + N_2(g).$$

The results are shown below.

t / s	0	1000	2000	3000	4000
$p/N\ m^{-2}$	10.8	7.5	5.4	3.9	2.6

(a) What is the rate equation for this first order decomposition process? What is the integrated form of the rate law?
(b) Plot a suitable graph to verify that this decomposition is first order, and use it to determine the rate constant for the decomposition process at 550 K.
(c) What is the half-life of azomethane at 550 K?

Note that the total pressure within the reaction vessel will be the sum of the partial pressures of azomethane, ethane and nitrogen.

Answers

(a) If the reaction is first order then the rate law for the decay of azomethane is:

$$\frac{d\{p(\text{azomethane})\}}{dt} = -k_1\{p(\text{azomethane})\}$$

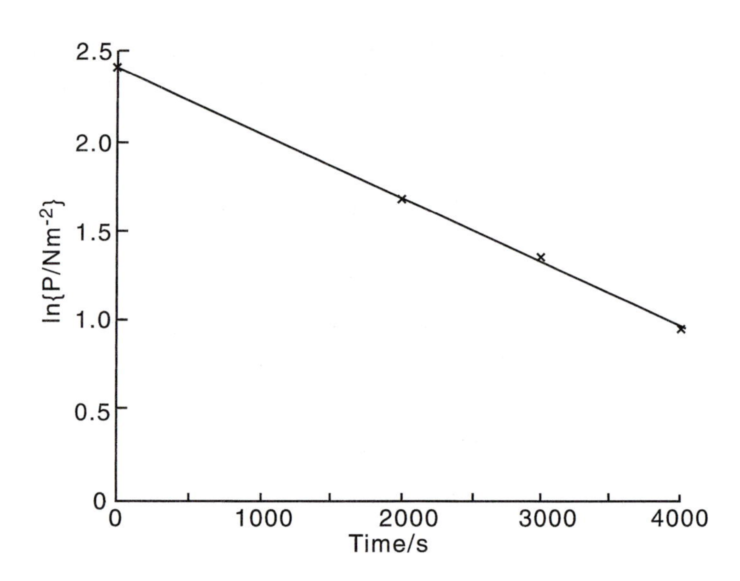

Fig. 4.10 A first order plot for the data in Question 11.

Consequently the integrated form is

$$\ln \{p(\text{azomethane})\} = -k_1 t + \text{constant}$$

(b) A graph of ln {p(azomethane)} against t, will be linear if the kinetic assumption is correct. Fig. 4.10 shows the data produced plotted in the required form. The graph is indeed linear so verifying the presumption of first order kinetics. The slope of the plot is $-0.0004\ s^{-1}$ so that,

$$k_1 = 0.0004\ s^{-1}.$$

(c) For a first order process the half-life is independent of concentration and given by

$$t_{\frac{1}{2}} = \frac{\ln 2}{k_1}$$

so that

$$t_{\frac{1}{2}} = 2000\ s.$$

Inspection of the data given shows that this is a realistic estimate; the pressure of azomethane does indeed fall by a factor of $\frac{1}{2}$ for every 2000 seconds from the start of the reaction.

Question 12: Integrated rate laws: second order kinetics

The reaction $CH_3CH_2NO_2 + OH^- \rightarrow CH_3CHNO_2^- + H_2O$ was carried out at 293 K with equal concentrations of each reactant. The OH^- concentration was found to vary as a function of time, t, as follows.

t/min	0	5	10	15
$10^3 \times [OH^-]$ /mol dm^{-3}	5.0	2.6	1.7	1.3

Show graphically that these results are consistent with those predicted for second order kinetics. Hence, determine the second order rate constant.

Answer

If the reaction is second order then, its rate law will be

$$R = \frac{d[OH^-]}{dt} = -k_2[OH^-][CH_3CH_2NO_2].$$

Since the concentration of OH^- and $CH_3CH_2NO_2$ are equal at the start of the reaction and must remain so throughout its course due to the one to one reaction stoichiometry, it follows that

$$\frac{d[OH^-]}{dt} = -k_2[OH^-]^2$$

so that

$$\frac{1}{[OH^-]} - \frac{1}{[OH^-]_0} = k_2 t.$$

Fig. 4.11 shows a plot of $[OH]^{-1}$ against time. This is linear verifying the hypothesis of second order kinetics. The slope of the graph has a value of 38 $mol^{-1}\ dm^3\ s^{-1}$ which corresponds to the value of k_2.

Question 13

The atmosphere contains three isotopes of carbon: ^{12}C, ^{13}C, ^{14}C: the latter is radioactive. Plants and animals usually do not distinguish between these isotopes. Consequently, the relative concentration of ^{14}C as compared to ^{12}C is

The relative abundance of the three naturally occurring isotopes of carbon are as follows:

^{12}C	^{13}C	^{14}C
98.9%	1.10%	trace

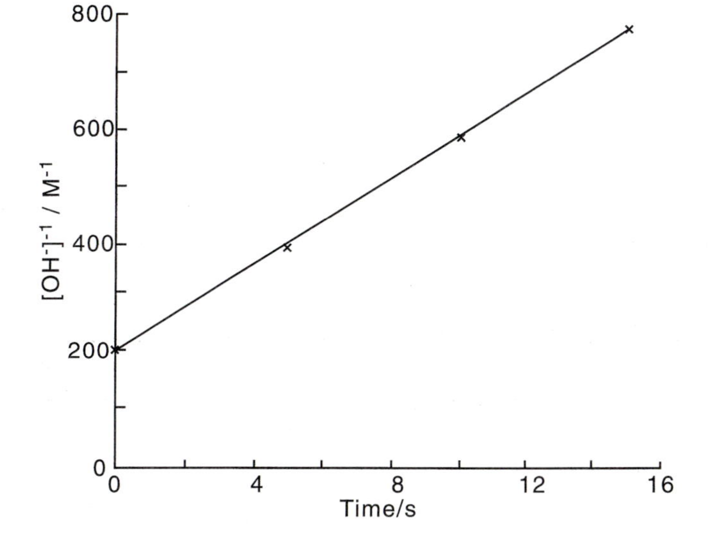

Fig. 4.11 A graph of the data in Question 12 showing second order kinetics apply.

^{14}C has been used widely as a dating technique. The main source of natural ^{14}C is the upper atmosphere where it is formed by the interaction of cosmic ray neutrons with ^{14}N at a rate of 10^2 ^{14}C atoms per cm^2 per minute. Natural levels have been diluted by combustion of fossil fuels—which contain no ^{14}C—but enhanced by explosions of nuclear bombs. See H. E. Svess, *Science*, 122, (1955), 415, and M. S. Baxton and A. Walton, *Proc. Royal, Soc. London*, A321, (1971), 105. ^{14}C is a β emitter; the loss of an electron from its nucleus regenerates ^{14}N.

kept constant during the lifetime of a living species. However, when a plant or animal dies, the ^{14}C isotope decays with a half-life of 5730 years.

An archaeological hunt in Africa revealed some preserved bark that was found to have 72% of the ^{14}C found in present day trees. What was the age of the fossil?

Answer

If the decay ^{14}C is first order the concentration of ^{14}C is given by

$$[^{14}C] = [^{14}C]_0 \exp(-k_1 t) \tag{4.18}$$

where $[^{14}C]_0$ is the concentration of ^{14}C found in living trees. The first order rate constant k_1 can be found from the relationship for the half life:

$$t_{\frac{1}{2}} = \frac{\ln 2}{k_1}$$

so that

$$k_1 = \frac{\ln 2}{5730} = 1.2 \times 10^{-4} \text{ years}^{-1}.$$

In the specimen,

$$\frac{[^{14}C]}{[^{14}C]_0} = 0.72$$

so that from equation (4.18)

$$0.72 = \exp(-t^* \times 1.2 \times 10^{-4}) \tag{4.19}$$

where t^* is the age of the bark in years. Solving equation (4.19) gives

$$t^* = 2716 \text{ years}.$$

Question 14: Kinetics in the gas phase as revealed by measurements of total pressure

(a) A gas phase decomposition reaction was followed by sealing the reactant in a vessel and following the total pressure, P, as a function of time, t. The following results were obtained.

t/min	0	1	3	6	12	40
P/atm	1.000	1.221	1.528	1.777	1.950	2.000

Show that the reaction is first order and find its half-life and rate constant.

(b) The thermal decomposition of dimethyl ether follows the equation

$$(CH_3)_2O(g) \rightarrow CH_4(g) + H_2(g) + CO(g)$$

For initial pressures, $P_0 = 4.16 \times 10^4$ N m^{-2} at a temperature of 777 K and $P_0 = 5.60 \times 10^4$ N m^{-2} at 825 K, the following pressure increases, Z, were measured as a function of time, t.

T = 777 K		T = 825 K	
t/s	$10^3 Z$ / N m^{-2}	t / s	10^3 Z / N m^{-2}
770	23.5	114	43.1
1195	33.3	219	71.2
3155	62.3	405	98.6
∞	83.2	∞	112.0

(i) Show how the pressure increase during the course of the above reaction is related to the fraction of $(CH_3)_2O$ which has decomposed.
(ii) Show that the reaction is first order.
(iii) Determine the rate constants at 777 K and 825 K and the activation energy.

Answers

(a) Interpretation of the data shows that if the initial pressure is compared to that reached after a long time then it can be seen that the total pressure doubles in the course of the reaction. The stoichiometry of the process is therefore of the form,

$$A(s) \rightarrow 2B(g).$$

If the initial pressure is P_o corresponding to pure A, the final pressure is $2P_o$ corresponding to pure B. This transition can be summarised by the following table in which the total pressure, P_{TOT}, is the sum of the partial pressures of A and B.

Time/min	Partial pressure of A (p_A)/atm	Partial pressure of B (p_B)/atm	Total pressure (P_{TOT})/atm
0	P_o	0	P_o
t	$(1-x)P_o$	$2xP_o$	$(1+x)P_o$
∞	0	$2P_o$	$2P_o$

In the table x represents the fraction of A converted into B.
To conduct the kinetic analysis we need to know p_A as a function of time. It may be deduced that

$$p_A = 2P_o - P_{TOT}$$

So that the partial pressure of A can be found from the final pressure ($2P_o = 2$ atmospheres) and the pressure reported at other times (P_{TOT}). If the reaction is first order a plot of ln p_A against time will be linear. Fig 4.12 shows $\ln(2 - P_{TOT})$ plotted against time; the graph is linear confirming first order kinetics and the slope is,

$$-k_1 = -0.25 \text{ min}^{-1}$$

so that

$$k_1 = 0.25 \text{ min}^{-1}.$$

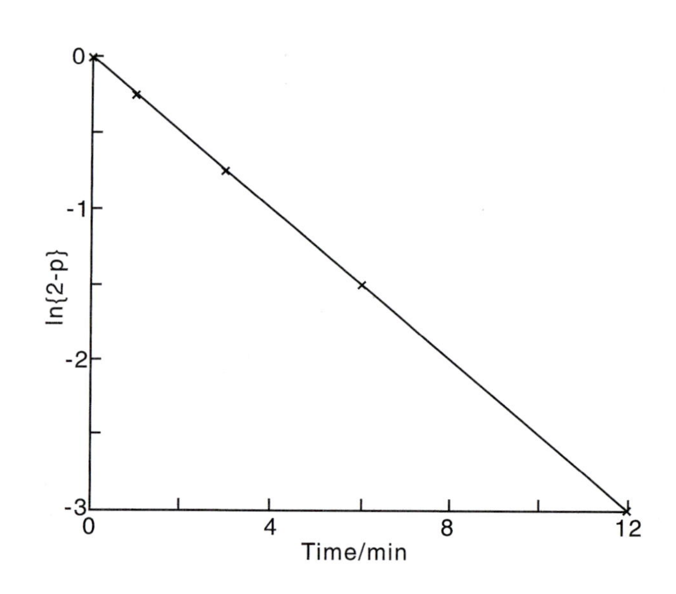

Fig. 4.12 A graph to show the data in Question 14 (a) follows first order kinetics.

(b) (i) The initial pressure, P_o, corresponds to pure $(CH_3)_2O$ whereas the final value corresponds to a value of $3P_o$ since the stoichiometry of the reaction shows that one mole of $(CH_3)_2O$ forms three moles of gaseous products. The table of data shows that the final *increase*,

$$Z_\infty = 2P_o.$$

At other times

$$P_{TOT} = p((CH_3)_2O) + p(CH_3) + p(H_2) + p(CO)$$
$$P_{TOT} = p((CH_3)_2O) + 3p(H_2),$$

since equal amounts of CH_4, H_2 and CO_2 are formed. Moreover, the number of moles of ether reacted must equal the number of moles of H_2 formed:

$$P_o - p((CH_3)_2O) = p(H_2)$$
$$P_o = p((CH_3)_2O) + p(H_2)$$

The pressure increase,

$$Z = P_{TOT} - P_o$$
$$= 2p(H_2)$$

If the reaction is first order in $p((CH_3)_2O)$ then

$$p((CH_3)_2O) = P_o \exp(-k_1 t)$$

where k_1 is the first order rate constant, but

$$p((CH_3)_2O) = p(H_2) - P_o$$
$$= \frac{Z_\infty - Z}{2}$$

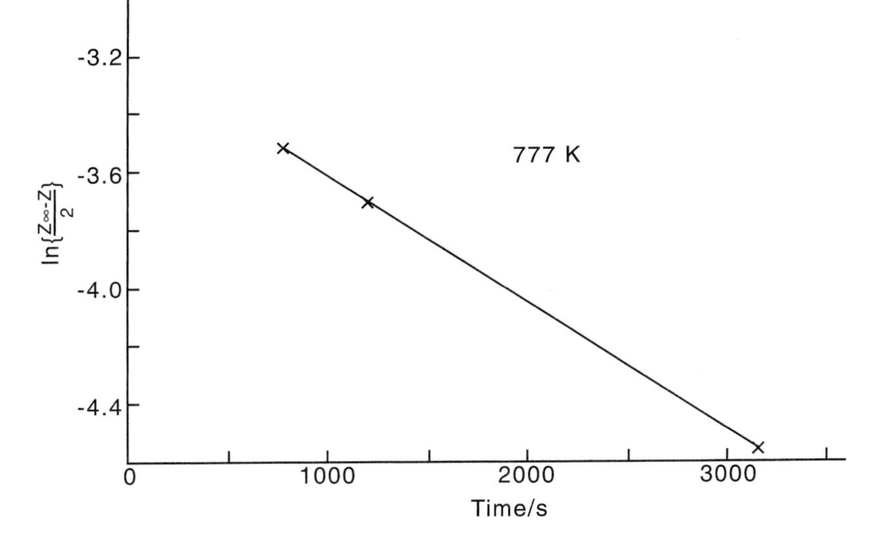

Fig. 4.13 A plot to determine the rate constant at 777 K.

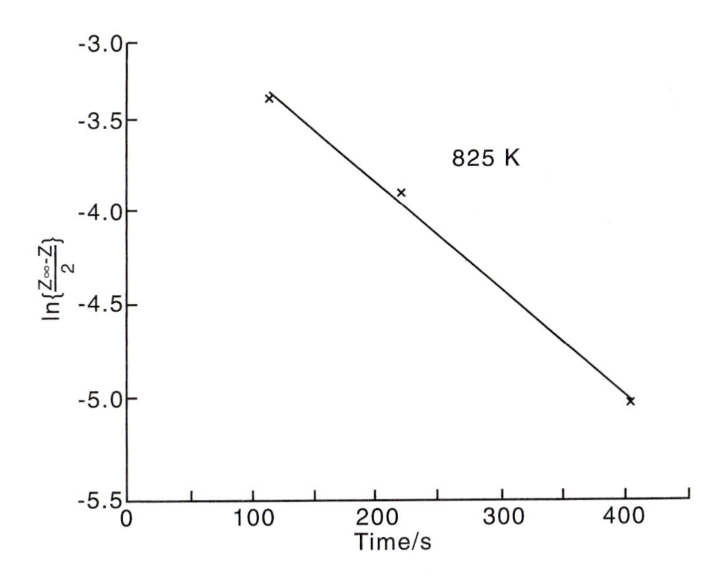

Fig. 4.14 A plot to determine the rate constant at 825 K.

(ii) Fig. 4.13 and Fig. 4.14 therefore shows plots of $\ln\left(\frac{Z_\infty - Z}{2}\right)$ against time at 777 K and 825 K. These are linear confirming first order kinetics apply.

(iii) The rate constants for each temperature can be found using the graphs in Fig. 4.13 and Fig. 4.14. Each linear graph gives a gradient of $-k_1$. Therefore the rate constant at 777 K is:

$$k_1 = 0.0004\ s^{-1}.$$

The rate constant at 825 K is:

$$k_1 = 0.0057\ s^{-1}.$$

Using the Arrhenius equation we can determine the activation energy. We know that

$$\ln k = \ln A - E_a/RT. \tag{4.20}$$

Substituting the data into equation (4.20), we find

$$\ln(0.0057/0.0004) = -\frac{E_a}{8.31}\left(\frac{1}{825} - \frac{1}{777}\right).$$

This gives an activation energy of

$$E_a = 295\ kJ\ mol^{-1}.$$

Question 15: The steady state approximation

(a) Explain what is meant by the steady state approximation in chemical kinetics. Why is it useful and under what conditions is it valid?

(b) The following mechanism has been proposed for the thermal decomposition of gaseous NO_2Cl.

$$NO_2Cl \underset{k_{-1}}{\overset{k_1}{\rightleftharpoons}} NO_2 + Cl$$

$$NO_2Cl + Cl \xrightarrow{k_2} NO_2 + Cl_2$$

(i) Derive an expression for the rate of decomposition of NO_2Cl in terms of the concentrations of NO_2Cl and NO_2, and the three rate constants k_1, k_{-1} and k_2.

(ii) Discuss the conditions under which the overall reaction will exhibit kinetics that are (A) first order with respect to NO_2Cl and (B) second order with respect to NO_2Cl.

(c) Hydrocarbon oxidation by dissolved oxygen is thought to occur via the following free-radical mechanism.

$$RH + O_2 \xrightarrow{k_1} R^\bullet + {}^\bullet O_2H$$

$$R^\bullet + O_2 \xrightarrow{k_2} RO_2^\bullet$$

$$RO_2^\bullet + RH \xrightarrow{k_3} RO_2H + R^\bullet$$

$$2R^\bullet \xrightarrow{k_4} R_2$$

Hint: the $HO_2^\bullet$ radical is relatively unreactive and can therefore neglected in the calculations.

(i) Use the steady state approximation applied to $RO_2^\bullet$ to find an expression relating k_2 and k_3.

(ii) Apply the steady state approximation to $R^\bullet$ to find an expression relating k_1, k_2, k_3, and k_4.

(iii) Hence derive an expression for $[R^\bullet]$, and deduce the rate of reaction, $\frac{d[RO_2H]}{dt}$, expressed in terms of [RH] and $[O_2]$.

Answers

(a) The steady state approximation applies to reaction intermediates, int, in complex reaction mechanisms for which

$$\frac{d[\text{int}]}{dt} \approx 0.$$

This implies that the species is highly unstable and rapidly decomposes so that only a low, but near constant concentration is built up. The use of the steady state approximation is to simplify complex reaction schemes as will be shown in parts (b) and (c) below.

(b) (i) The reaction given proceeds through the unstable intermediate, Cl, to which we apply the steady state approximation. If

$$\frac{d[Cl]}{dt} \approx 0$$

then the rate at which Cl atoms are made must exactly balance the rate at which they are destroyed:

Rate of making Cl = Rate of loss of Cl,

so that

$$k_1[NO_2Cl] = k_{-1}[Cl][NO_2] + k_2[Cl][NO_2Cl].$$

It follows that the steady state concentration is simply:

$$[Cl]_{ss} = \frac{k_1[NO_2Cl]}{k_{-1}[NO_2] + k_2[NO_2Cl]}.$$

The sought rate is the loss of NO_2Cl

$$\begin{aligned}\frac{d[NO_2Cl]}{dt} &= -k_1[NO_2Cl] + k_{-1}[Cl]_{ss}[NO_2] - k_2[Cl]_{ss}[NO_2Cl] \\ &= -k_1[NO_2Cl] - \frac{k_1[NO_2Cl](k_2[NO_2Cl] - k_{-1}[NO_2])}{k_{-1}[NO_2] + k_2[NO_2Cl]} \\ &= \frac{-2k_1k_2[NO_2Cl]^2}{k_{-1}[NO_2] + k_2[NO_2Cl]}.\end{aligned}$$

(ii) (A) The reaction will be first order with respect to $[NO_2Cl]$ if

$$k_{-1}[NO_2] < k_2[NO_2Cl]$$

so that

$$\frac{d[NO_2Cl]}{dt} = -2k_1[NO_2Cl]$$

In this limit the reaction sequence is

$$NO_2Cl \xrightarrow{\text{slow}} NO_2 + Cl$$
$$NO_2Cl + Cl \xrightarrow{\text{fast}} NO_2 + Cl_2$$

(B) The reaction will be second order if

$$k_{-1}[NO_2] > k_2[NO_2Cl]$$

so that

$$\frac{d[NO_2Cl]}{dt} = \frac{-2k_1k_2[NO_2Cl]^2}{k_{-1}[NO_2]}.$$

This limit may be associated with the reaction sequence:

$$NO_2Cl \rightleftharpoons NO_2 + Cl$$

$$NO_2Cl + Cl \rightarrow NO_2 + Cl_2$$

The rate of the second step is,

$$\frac{d[NO_2Cl]}{dt} = -k_2[NO_2Cl][Cl]$$

and
$$\frac{k_1}{k_{-1}} = \frac{[NO_2][Cl]}{[NO_2Cl]} = K_{eq}$$

where K_{eq} is the equilibrium constant for the first step. Combining these equations,

$$\frac{d[NO_2Cl]}{dt} = -k_2K_{eq}\frac{[NO_2Cl]^2}{[NO_2]}.$$

(c) (i) Applying the steady state approximation to $RO_2^{\bullet}$.

$$\frac{d[RO_2^{\bullet}]}{dt} = 0.$$

So that the rate of making $RO_2^{\bullet}$ must balance the rate of its destruction.

$$k_2[R^{\bullet}][O_2] = k_3[RH][RO_2^{\bullet}] \tag{4.21}$$

(ii) Likewise, if

$$\frac{d[R^{\bullet}]}{dt} = 0$$

then the rate of formation and destruction of $R^{\bullet}$ must be equal:

$$k_1[RH][O_2] + k_3[RO_2^{\bullet}][RH] = k_2[R^{\bullet}][O_2] + k_4[R^{\bullet}]^2 \tag{4.22}$$

Subtracting equations (4.21) and (4.22) gives

$$k_1[RH][O_2] = k_4[R^{\bullet}]^2$$

so that

$$[R^\bullet] = \sqrt{\frac{k_1}{k_4}}[RH]^{\frac{1}{2}}[O_2]^{\frac{1}{2}}.$$

Using equation (4.21) we can deduce that

$$k_2\sqrt{\frac{k_1}{k_4}}[RH]^{\frac{1}{2}}[O_2]^{\frac{3}{2}} = k_3[RH][RO_2^\bullet]$$

$$= \frac{d[RO_2H]}{dt}.$$

Question 16: Not all reactions occur via activation!

Measurements have been made of the rates of the reactions

$$H^\bullet(g) + H_2(g) \rightarrow H_2(g) + H^\bullet(g)$$

and

$$D^\bullet(g) + D_2(g) \rightarrow D_2(g) + D^\bullet(g)$$

In particular, the second order rate constants, k, have been measured as a function of temperature. Fig. 4.15 shows a plot of $\log_{10}$ k against reciprocal temperature for both systems.

(a) Explain the purpose in plotting $\log_{10}$ k against 1/T, and why the plots might be expected to be linear.

(b) Estimate the activation energy for the reactants using the high temperature data.

(c) The plot for the $H^\bullet + H_2$ reaction becomes curved at low temperatures: the reaction appears to go faster than it should on the basis of extrapolation of the high temperature data. Explain this phenomenon.

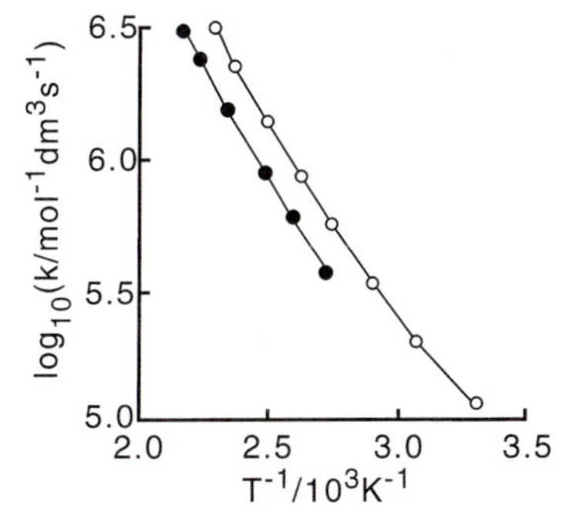

Fig. 4.15 Arrhenius plots for hydrogen exchange (○) and deuterium exchange (•) reactions.

Answer

(a) The Arrhenius equation predicts,

$$k = A \exp\left(-\frac{E_a}{RT}\right)$$

so that,

$$\ln k = \ln A - \frac{E_a}{RT}$$

or

$$\log_{10}k = \log_{10}A - \frac{E_a}{2.3RT}$$

It follows that a plot of $\log_{10}$k against 1/T will be linear if the Arrhenius equation is applicable to the data and, if so, the slope will permit the activation energy to be deduced.

(b) From Fig. 4.15 , the slope of the linear part of the graph is approximately -1.98×10^3 K so that E_a is 37 kJ mol^{-1}.

(c) Extrapolation of the high temperature data for the $H^\bullet + H_2$ system predicts much lower rate constants than are actually observed. These

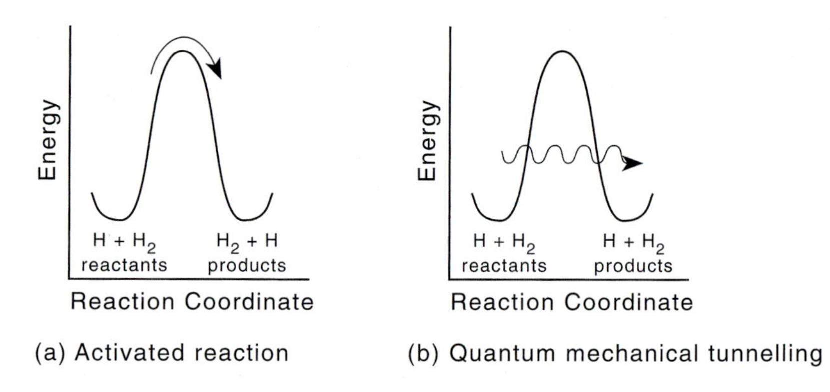

Fig. 4.16 Reaction profiles showing how reactants form products (a) classically (with Arrhenius type behaviour), and (b) quantum mechanically (with non-Arrhenius type behaviour).

estimates are based on the Arrhenius equation which presumes that reaction occurs by collisions in the gas phase providing enough energy—the activation energy—to cross the energy barrier for the reaction, as shown in Fig. 4.16(a). However, in the case of light particles—especially electrons but also hydrogen atoms—quantum mechanics predicts, correctly, that there is a small but finite probability of passage *through* rather than over the energy barrier, Fig. 4.16(b)! That is to say, in the present case, the reactants become products without ever having the energy required to surmount the energy barrier! This behaviour is, of course, entirely alien to our expectations based on the macroscopic world around us. Nevertheless, quantum theory predicts, and experiment confirms, the reality of this phenomenon of *tunnelling*. Other examples of this include:

The probability of tunnelling of atoms heavier than H atoms is almost negligible.

- the loss of β-particles (electrons) from radioactive nuclei,
- electron transfer from the surface of electrodes during electrolysis, where the electron tunnels between the electrodes and the solution phase species so that it is never "free" in solution, and
- homogeneous electron transfer in solution, for example the reaction

$$Fe^{3+}(aq) + I^{-}(aq) \rightarrow Fe^{2+}(aq) + I^{\bullet}.$$

Here the electron tunnels from the iodide ion to the Fe^{3+} species.

Given that there is a tunnelling contribution to the $H^{\bullet} + H_2$ reaction we must in this case modify the expression for the rate constant,

$$k = A \exp\left(-\frac{E_a}{RT}\right) + B \tag{4.23}$$

where B is a largely temperature invariant parameter describing the tunnelling contribution. Equation (4.23) shows that the reaction between $H^{\bullet}$ and H_2 occurs *both* via activated reaction (passage over the barrier) and tunnelling (passage through the barrier). Inspection of equation (4.23) shows that at high temperatures the former term dominates, so conventional Arrhenius behaviour is seen, but at low temperatures, the activated path becomes infinitesimally slow and the quantum tunnelling becomes apparent:

$$k \approx A \exp\left(-\frac{E_a}{RT}\right) \qquad \text{high T}$$

$$k \approx B \qquad \text{low T.}$$

Question 17

(a) Using the Arrhenius equation, derive an equation describing the variation of the vapour pressure of a liquid as a function of temperature.
(b) The temperature of the vapour pressure of CO_2 is as follows:

T / K	180	190	200	210	220	230	240	250
p / atm	0.277	0.687	1.553	3.274	5.962	8.876	12.74	17.71

Draw a suitable graph to estimate:
(i) the triple point,
(ii) the enthalpy change on sublimation, and
(iii) the enthalpy change on vaporisation.

Answer

(a) We consider the following *equilibrium*,

$$A(\ell) \rightleftharpoons A(g)$$

and develop an equation for the vapour pressure, p_A of A as a function of temperature, T. The fact that the system is at equilibrium means that the rate at which A molecules are leaving the liquid to enter the gas phase must be exactly balanced by the rate at which gaseous A molecules return to the liquid. That is, a *dynamic* equilibrium pertains.

The rate of evaporation will depend on the temperature, the surface area of the liquid and the chemical identity of the latter. We use the rate constant k_e to quantify this rate which will be measured in units of mol s^{-1}. Next, we turn to the rate of condensation. We again expect this to depend on the same three parameters as for the rate of evaporation, but also the amount of A present in the gas phase as reflected by its partial pressure, p_A. The rate of condensation is therefore given by the product of a rate constant k_c and p_A. At equilibrium,

$$\text{rate of evaporation} = \text{rate of condensation}$$

so,

$$k_e = k_c p_A \qquad (4.24)$$

and

$$\ln k_e = \ln k_c + \ln p_A. \qquad (4.25)$$

Now suppose that both k_e and k_c can be described by the Arrhenius equation:

$$k_e = A_e \exp\left(-\frac{E_a^e}{RT}\right) \tag{4.26}$$

$$k_c = A_c \exp\left(-\frac{E_a^c}{RT}\right). \tag{4.27}$$

The two activation energies, E_a^e and E_a^c are illustrated in Fig. 4.17. Substituting equations (4.26) and (4.27) into equation (4.25), we see that

$$\ln A_e - \frac{E_a^e}{RT} = \ln A_c - \frac{E_a^c}{RT} + \ln p_A \tag{4.28}$$

or,

$$\ln p_A = -\frac{(E_a^e - E_a^c)}{RT} + \text{constant} \tag{4.29}$$

where the constant depends on the pre-exponential factors A_e and A_c. The first term on the right hand side of equation (4.29) contains the difference of the two activation energies, $(E_a^e - E_a^c)$. Inspection of Fig. 4.17 shows that this is simply the energy difference between $A(\ell)$ and $A(g)$: that is the enthalpy of vaporisation of A, ΔH°_{vap}. Equation (4.29) can now be simplified:

$$\ln p_A = -\frac{\Delta H^\circ_{vap}}{RT} + \text{constant.} \tag{4.30}$$

This is the well-known Clausius-Clapeyron equation. It tells us that a plot of ln p_A against $1 / T$ should be linear with a slope of $(-\Delta H^\circ_{vap}/R)$. A little thought will show that if a similar argument to the above is applied to sublimation,

$$A(s) \rightleftharpoons A(g)$$

The derivation of the Clausius-Clapeyron equation presented here is a kinetically based argument. It can, in fact, be derived solely on the basis of rigorous thermodynamic arguments. See G. Price, "Thermodynamics of Chemical Processes", (OCP 56) for details.

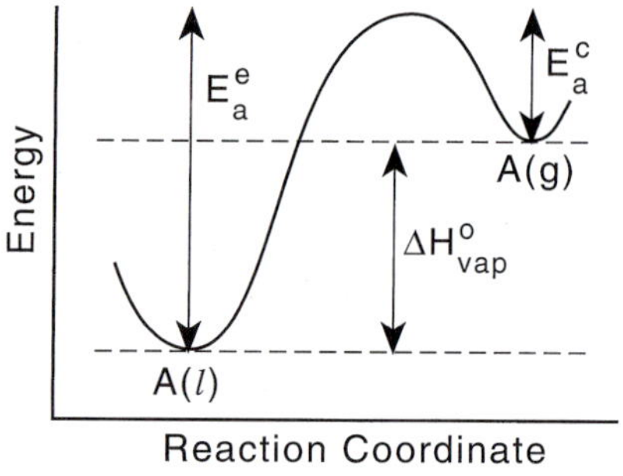

then the vapour pressure of a gas above a solid, such as dry ice (solid carbon dioxide) will be given by

$$\ln p_A = -\frac{\Delta H^\circ_{sub}}{RT} + \text{constant} \tag{4.31}$$

where ΔH°_{sub} is the enthalpy of sublimation.

Fig. 4.17 Reaction profile for vaporisation.

(b) Answering, finally, the question posed in equation (4.31) suggests a suitable plot is that of ln p against (1/T). Such a graph for the data gives relating to CO_2 is plotted in Fig. 4.18.

It can be seen that the plot comprises of two linear portions. At low temperature the data relates to the sublimation:

$$CO_2(s) \rightleftharpoons CO_2(g)$$

whereas at the high temperature and pressure the CO_2 is liquid so that vaporisation occurs:

$$CO_2(\ell) \rightleftharpoons CO_2(g).$$

(i) At the triple point the gas, liquid and solid phase co-exist. Since the two lines in Fig. 4.18 refer to the gas/solid and gas/liquid equilibrium it follows that the only point where all three phases co-exist is the point of intersection of the lines. This corresponds to a value of 5.11 atm and a temperature of 217 K.

(ii) The slope of the line for low temperature data (large values of 1/T) is −3103 K. Considering equation (4.31) shows that

$$-\frac{\Delta H^{\circ}_{sub}}{R} = -3103 \text{ K}$$

so that $$\Delta H^{\circ}_{sub} = 26 \text{ kJ mol}^{-1}.$$

(iii) The slope of the line in Fig. 4.18 shown for the high temperature data (lower values of 1/T) is −2181 K. It follows from equation (4.31) that

$$-\frac{\Delta H^{\circ}_{vap}}{R} = -2181 \text{ K}$$

so that $\Delta H^{\circ}_{vap} = 18 \text{ kJ mol}^{-1}$.

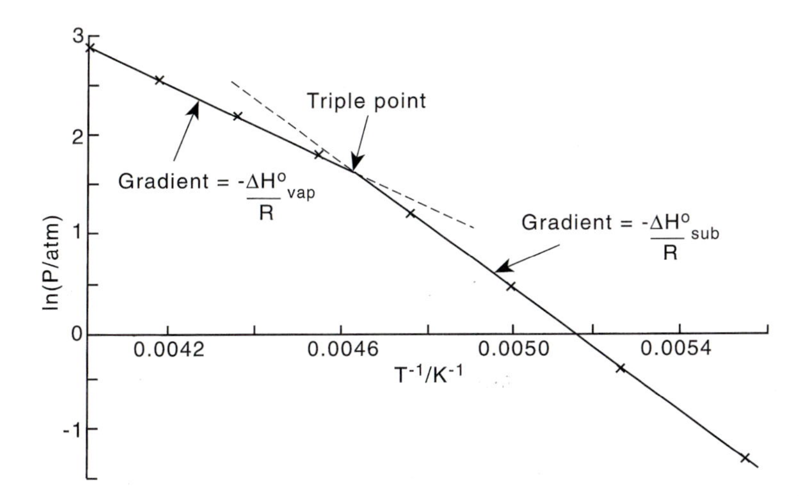

Fig. 4.18 Clausius-Clapeyron plot for Question 17.

5 Chemical equilibria

5.1 Aims

This chapter seeks to provide a range of problems to illuminate the following topics.

- Le Chatelier's principle.
- Equilibrium constants and chemical speciation in both gases and solutions.
- The relationship between the standard Gibbs free energy changes of a reaction, ΔG°, and the equilibrium constant for the reaction.
- Acid-base equilibria.
- Complex-ion formation.
- Solubility of solids.
- Redox equilibria.

Problems concerning core knowledge in these areas are presented together with questions that move in the direction of more advanced material.

5.2 Le Chatelier's principle

Question 1

(a) State Le Chatelier's principle.

(b) Predict the changes in the following equilibrium that occur when:

(i) the temperature is raised, and

(ii) the pressure is increased.

$$CH_4(g) + 2O_2(g) \rightleftharpoons CO_2(g) + 2H_2O(\ell).$$

The following data will be useful.

	ΔH°_f/ kJ mol^{-1}
$H_2O(\ell)$	–286
$CH_4(g)$	–75
$CO_2(g)$	–394

Answer

(a) Le Chatelier's principle is that: *if a system in equilibrium is exposed to a change, the equilibrium shifts to oppose that change.*

(b) (i) The data given shows that for the reaction, ΔH° is –891 kJ mol^{-1}. The process is exothermic so that the equilibrium will shift in favour of the reactants, CH_4 and O_2, as the temperature is raised.

(ii) The reaction involves the conversion of three moles of gas into one mole of gas and two of liquid water. Increasing the pressure will therefore favour the products since they will be formed with a reduction in volume or pressure.

Question 2: Le Chatelier's principle: explaining the effect of pressure

For the equilibrium

$$CH_4(g) \rightleftharpoons 2H_2(g) + C(s,\ graphite),$$

find an expression relating the equilibrium constant, K_p, to the degree of dissociation of methane, α, and the total pressure, P_{TOT}. Hence, predict how compression would affect the mole fraction of CH_4 in an equilibrium mixture of CH_4 and H_2, and graphite. Comment, in the light of Le Chatelier's principle.

Answer

Suppose that before the reaction takes place there are n moles of CH_4. After reaction, a fraction α, has decomposed so that there will be $n(1-\alpha)$ moles of CH_4 remaining. At the same time, $n\alpha$ moles of solid graphite will be formed along with $2n\alpha$ moles of H_2. It is often helpful to summarise this type of information beneath the equation for the reaction:

	$CH_4(g)$	$\rightleftharpoons$	$2H_2(g)$	+	C(s, graphite)
Before reaction:	n moles		0 moles		0 moles
After reaction:	$n(1-\alpha)$ moles		$2n\alpha$ moles		$n\alpha$ moles

The equilibrium constant, K_p, for the reaction is given by

$$K_p = \frac{p(H_2(g))^2}{p(CH_4(g))}.$$

Notice there is no mention of the graphite since this is a solid. The partial pressures of hydrogen, $p(H_2)$ and methane, $p(CH_4)$ are related to the total pressure, P_{TOT}, in the equations

$$p(H_2) = x(H_2)\ P_{TOT}$$
$$p(CH_4) = x(CH_4)\ P_{TOT}$$

where x denotes the relevant mole fraction in the gas phase when the equilibrium composition pertains. These are given by

$$x(H_2) = \frac{2n\alpha}{n(1-\alpha)+2n\alpha} = \frac{2\alpha}{1+\alpha}$$

and

$$x(CH_4) = \frac{n(1-\alpha)}{n(1-\alpha)+2n\alpha} = \frac{1-\alpha}{1+\alpha}.$$

Accordingly,

$$K_p = \frac{4\alpha^2 P_{TOT}/(1+\alpha)^2}{(1-\alpha)P_{TOT}/(1+\alpha)} = \frac{4\alpha^2 P_{TOT}}{1-\alpha^2}.$$

Note that the term $\frac{4\alpha^2}{1-\alpha}$ increases sharply as α changes from 0 to 1.

It follows that if the total pressure, P_{TOT}, is increased via compression, α must decrease so that K_p remains constant. This behaviour is exactly that predicted qualitatively by Le Chatelier's principle.

Question 3: Le Chatelier's principle: explaining the effect of temperature

(a) The reaction

$$PCl_5(g) \rightleftharpoons PCl_3(g) + Cl_2(g)$$

has an equilibrium constant $K_p = 3.0 \times 10^{-7}$ atm at a temperature of 298 K. Would this value be higher or lower at a temperature of 500 K given the following standard enthalpies of formation: ΔH_f° (PCl_5), –375 kJ mol^{-1}; ΔH_f° (PCl_3), –320 kJ mol^{-1}?

(b) (i) By using *kinetic* arguments derive an expression for K_p in terms of the partial pressures of PCl_5 and PCl_3 and Cl_2.

(ii) Derive an expression relating K_p to the standard enthalpy change for the reaction, ΔH°.

(c) Use the expression established in part (b)(i) to calculate a value of K_p at 500 °C.

Answer

(a) By applying a Hess cycle, the enthalpy change for the dissociation reaction, ΔH°, can be found to be +55 kJ mol^{-1}. The process is therefore endothermic—Le Chatelier's principle predicts that the formation of the products PCl_3 and Cl_2 will be favoured by an increase in temperature from 298 K to 500 K.

(b) (i) To apply kinetic arguments, we identify the rate constants k_f and k_b with the forward and backward reactions:

$$PCl_5(g) \underset{k_b}{\overset{k_f}{\rightleftarrows}} PCl_3(g) + Cl_2(g).$$

If the forward process is assumed first order, whilst the backward process second order, at equilibrium the rates of the two must balance so,

$$k_f\, p(PCl_5) = k_b\, p(PCl_3)p(Cl_2).$$

Rearranging,

$$\frac{k_f}{k_b} = \frac{p(PCl_3)p(Cl_2)}{p(PCl_5)} = K_p \tag{5.1}$$

where K_p is the equilibrium constant for the reaction.

(ii) It is reasonable to suppose that the rate constants k_f and k_b are described by the Arrhenius equation,

$$k_f = A_f \exp(-E_a^f / RT) \tag{5.2}$$

$$k_b = A_b \exp(-E_a^b / RT) \tag{5.3}$$

where the activation energies, E_a^f and E_a^b, in the two directions can be understood by reference to Fig. 5.1. Substituting equations (5.2)

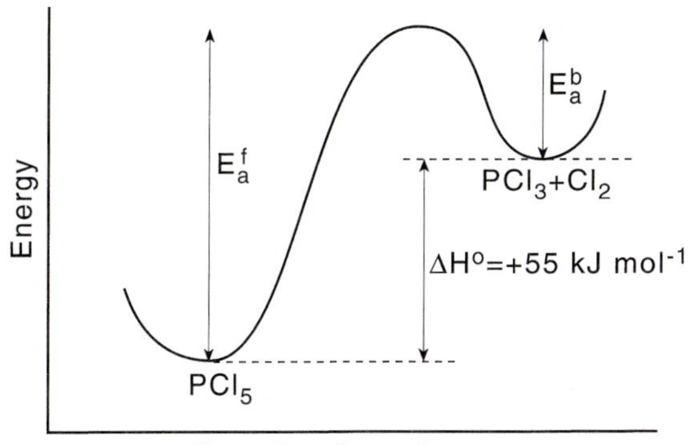

Fig. 5.1 Reaction profile for the reaction $PCl_5 \rightleftharpoons PCl_3 + Cl_2$.

and (5.3) into equation (5.1), it can be seen that

$$K_p = \frac{A_f}{A_b}\exp\left(-\frac{(E_a^f - E_a^b)}{RT}\right)$$

or

$$\ln K_p = \text{constant} - \frac{(E_a^f - E_a^b)}{RT}$$

$$= \text{constant} - \frac{\Delta H^\circ}{RT} \qquad (5.4)$$

where the constant is related to the pre-exponential factors A_f and A_b. Equation (5.4) is the famous van't Hoff Isochore which describes the temperature dependence of the equilibrium constant, K_p.

The van't Hoff Isochore can be derived rigorously from a pure thermodynamic standpoint without recourse to kinetic arguments. A derivation can be found in G. Price, "Thermodynamics of Chemical Processes", (OCP 56).

(c) Application of equation (5.4) to the reaction of interest above shows that at 550 °C (773 K),

$$\ln K_p(773) = \text{constant} - \frac{\Delta H^\circ}{773R}$$

whilst at 298 K,

$$\ln K_p(298) = \text{constant} - \frac{\Delta H^\circ}{298R}.$$

Solving gives:

$$\ln K_p(773) = \ln K_p(298) + \frac{\Delta H^\circ}{R}\left[\frac{1}{298} - \frac{1}{773}\right].$$

Substituting for $\Delta H^\circ = 55$ kJ mol^{-1} gives $K_p(773) = 0.25$ atm. This value is substantially greater than that found at 298 K.

Question 4: Finding ΔH° from measurements of K_p

Determine ΔH° at 1000 K for the reaction

$$I_2(g) \rightleftharpoons 2I(g)$$

from the following experimental measurements of K_p.

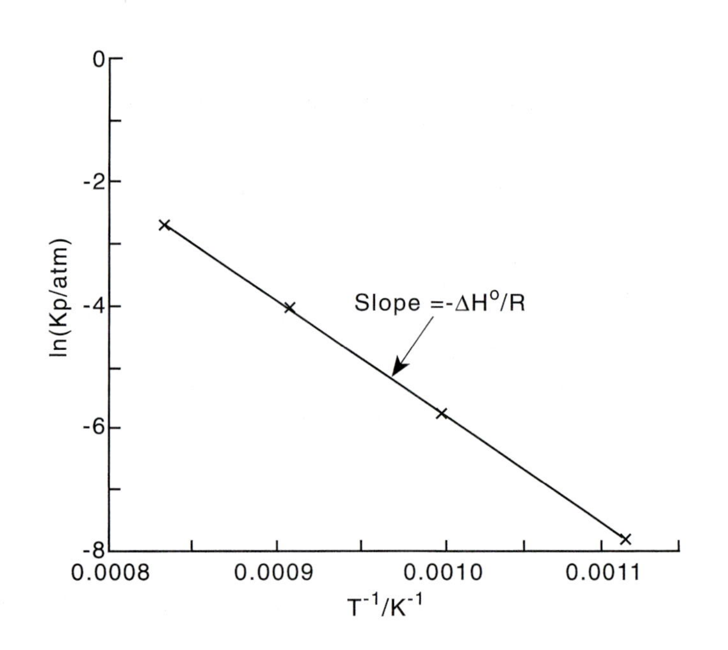

Fig. 5.2 Plot to determine the enthalpy of reaction from the temperature dependence of the equilibrium constant.

T / K	900	1000	1100	1200
K_p / atm	3.94×10^{-4}	3.03×10^{-3}	1.70×10^{-2}	6.72×10^{-2}

Answer

Equation (5.4) shows that for any reaction described by an equilibrium constant K_p.

$$\ln K_p = \text{constant} - \frac{\Delta H^\circ}{RT}$$

so that plotting a graph of ln K_p against $1/T$ should give a straight line of slope $-\frac{\Delta H^\circ}{R}$. Fig. 5.2 shows such a plot for the results of interest. It is a straight line and the slope,

In answering Question 4 it is assumed that ΔH° does not depend on temperature.

$$-\frac{\Delta H^\circ}{R} = -18555 \text{ K}$$

so that $\Delta H^\circ = +154$ kJ mol^{-1}.

Note that the reaction is endothermic as would be expected for a bond breaking reaction.

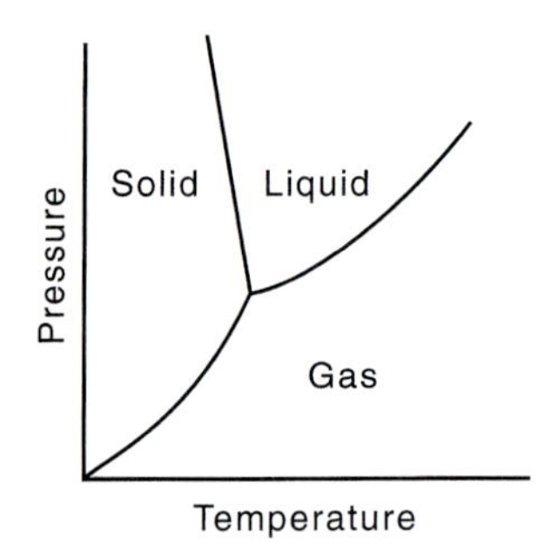

Fig. 5.3 Phase diagram for water.

5.3 Equilibrium constants and chemical speciation

Question 5: Phase equilibria: effects of pressure and temperature

Figs. 5.3 and 5.4 show the phase diagrams of water and carbon dioxide respectively. Comment on the variation of the phase boundaries with temperature and pressure.

Answer

In chapter four, the Clausius-Clapeyron equation, equation (4.30), was derived for the vapour pressure, p, of a liquid:

$$\ln p = -\frac{\Delta H^{\circ}_{vap}}{RT} + \text{constant} \qquad (5.5)$$

where ΔH°_{vap} is the enthalpy of vaporisation. Equation (5.5) describes the variation of pressure with temperature along the line which is the boundary between the liquid and gas phases in phase diagrams such as Figs. 5.3 and 5.4. It can be seen that since ΔH°_{vap} is always positive, p increases as T increases. This is exactly what would be predicted by applying Le Chatelier's principle to the vaporisation equilibrium of any species A,

$$A(\ell) \rightleftharpoons A(g) \qquad \Delta H^{\circ}_{vap} > 0.$$

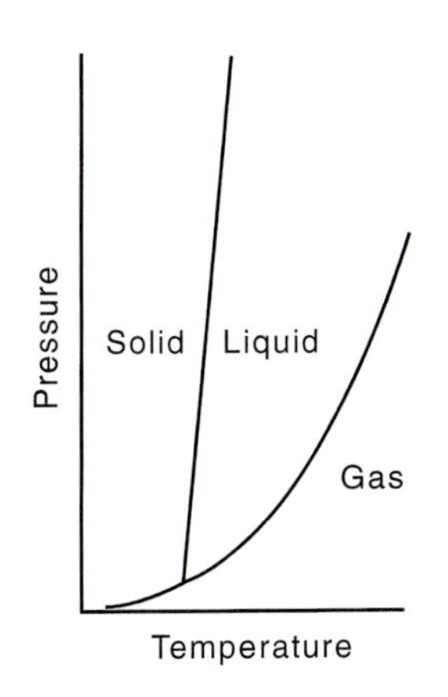

Fig. 5.4 Phase diagram for carbon dioxide.

Similarly for the solid-gas equilibrium,

$$A(s) \rightleftharpoons A(g) \qquad \Delta H^{\circ}_{vap} > 0$$

and again the equilibrium will be expected to shift in favour of the gaseous species as the temperature, as on the basis of Le Chatelier's principle, is varied. The corresponding equation to equation (5.5), is

$$\ln p = -\frac{\Delta H^{\circ}_{sub}}{RT} + \text{constant},$$

where ΔH°_{sub} is the enthalpy of sublimation. It is a general feature of phase diagrams that the slopes, $\frac{dp}{dT}$, of the boundary corresponding to gas-solid and gas-liquid equilibria are positive.

The equilibrium

$$A(s) \rightleftharpoons A(\ell) \qquad \Delta H^{\circ}_{fus} > 0$$

corresponding to the melting of A can, on inspection of Figs. 5.3 and 5.4, be seen to be different for CO_2 and for H_2O. In the first case, $\frac{dp}{dT}$ is positive, whereas in the latter it is negative. The former behaviour is more typical and corresponds to the usual case where the solid has a smaller value than the liquid. The two trends influencing the pressure-temperature behaviour of the melting point are:

- the endothermic nature of the melting process, so that higher temperature encourages the formation of the liquid phase, and
- the increase in volume on melting, so that an increased pressure favours the solid phase.

The two effects together ensure $\frac{dp}{dT} > 0$ for the solid-liquid phase boundary in CO_2, and in most other substances.

However, in the case of H_2O, the solid—ice—has a larger value than the liquid owing to its hydrogen bonded structure (Fig. 5.5). Accordingly, the effect of pressure is to encourage the formation of the liquid phase so that $\frac{dp}{dT} < 0$ for the liquid-solid phase boundary in H_2O.

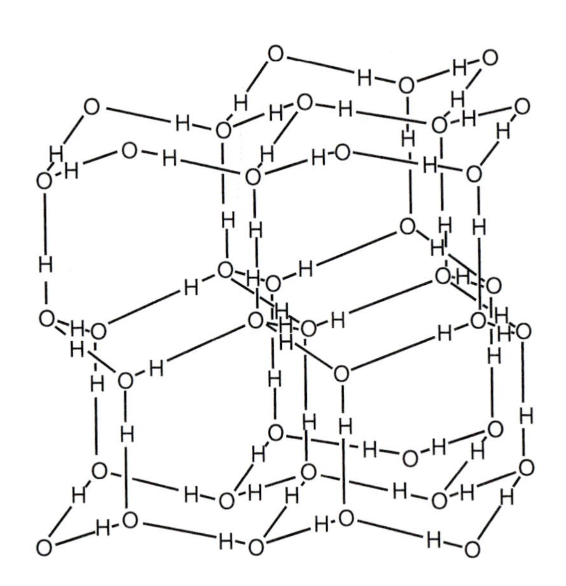

Fig. 5.5 The hydrogen bonded, tetrahedrally co-ordinated structure of ice.

5.4 Calculations on the equilibrium state

Question 6

When ethanol and ethanoic acid react together in the presence of a catalytic amount of sulphuric acid, an equilibrium is set up according to the equation

$$CH_3COOH(\ell) + C_2H_5OH(\ell) \rightleftharpoons CH_3COOC_2H_5(\ell) + H_2O(\ell).$$

(a) Write an expression for the equilibrium constant, K_c, expressed in terms of the concentration of the species involved.

(b) Given that $K_c = 4.0$ at 298 K, calculate the amount of ethyl ethanoate present at equilibrium, when two moles of CH_3COOH are equilibrated with two moles of C_2H_5OH.

Answer

(a) The equilibrium constant is given by

$$K_c = \frac{[CH_3COOC_2H_5(\ell)][H_2O(\ell)]}{[CH_3COOH(\ell)][C_2H_5OH(\ell)]}.$$

(b) It is helpful to adopt the strategy in Question 2 and keep track of the number of moles of reactant and products before reaction and at equilibrium

	CH_3COOH +	C_2H_5OH	$\rightleftharpoons$ $CH_3COOC_2H_5$	+ H_2O
Before reaction:	2	2	0	0
At equilibrium:	(2 − x)	(2 − x)	x	x

where x is the number of moles of ester found at equilibrium. The equilibrium constant K_c

$$K_c = \frac{(x/V)(x/V)}{((2-x)/V)^2} = \frac{(x)^2}{(2-x)^2} = 4 \qquad (5.6)$$

where the V is the volume of the mixture. Note the latter does not arise in the final expression for K_c.

Rearranging equation (5.6)

$$x^2 = 4(2-x)^2$$
$$= 4(4 - 4x + x^2)$$

or

$$3x^2 - 16x + 16 = 0$$

Solving the quadratic equation shows that the number of moles of ester in the product at equilibrium is

$$x = 1.33 \text{ or } 4$$

The first answer is correct since the second is clearly physically impossible.

The general solution of the quadratic equation $ax^2 + bx + c = 0$ is

$$x = \frac{-b \pm \sqrt{b^2 - 4ac}}{2a}.$$

Question 7

Nitrogen dioxide exists in equilibrium with its dimer (dinitrogen tetroxide):

$$2NO_2(g) \rightleftharpoons N_2O_4(g).$$

(a) Write an expression for the equilibrium constant, K_p, for this process.

(b) Explain qualitatively what would happen if the pressure of an equilibrium mixture of nitrogen dioxide and dinitrogen tetroxide were halved, whilst the temperature maintained constant, and the system left until the equilibrium re-established?

(c) The mole fraction of dinitrogen tetroxide in an equilibrium mixture with nitrogen dioxide was measured as being 0.23. The total pressure of the equilibrated system was 2.0×10^5 N m^{-2}. Calculate a value of K_p for this system.

Answer

(a) The equilibrium constant

$$K_p = \frac{p(N_2O_4)}{p(NO_2)^2}$$

where $p(N_2O_4)$ and $p(NO_2)$ are the partial pressures of N_2O_4 and NO_2 respectively.

(b) If the pressure is reduced Le Chatelier's principle tells us that the equilibrium will shift to oppose the change—that is in the direction to increase the pressure in the present instance. The equilibrium will therefore shift in the favour of the reactants since the overall increase in the number of moles of gas would tend to enhance the pressure.

(c) The mole fraction of N_2O_4 in the mixture is defined as

$$x(N_2O_4) = \frac{\text{no. of moles of } N_2O_4}{\text{no. of moles of } NO_2 + \text{no. of moles of } N_2O_4}.$$

Likewise, the mole fraction of NO_2 is similarly

$$x(NO_2) = \frac{\text{no. of moles of } NO_2}{\text{no. of moles of } NO_2 + \text{no. of moles of } N_2O_4}.$$

Clearly,

$$x(N_2O_4) + x(NO_2) = 1$$

so that if $x(N_2O_4) = 0.23$ then $x(NO_2) = 0.77$,
If the total pressure, P_{TOT} is 2×10^5 Nm^{-2} then the partial pressure of the two species are

$$p(NO_2) = x(NO_2)\, P_{TOT}$$
$$= 0.77 \times 2.0 \times 10^5 \text{ N m}^{-2}$$
$$= 1.54 \times 10^5 \text{ N m}^{-2}.$$

Likewise

$$p(N_2O_4) = x(N_2O_4) P_{TOT}$$
$$= 0.23 \times 2.0 \times 10^5 \text{ N m}^{-2}$$
$$= 4.6 \times 10^4 \text{ N m}^{-2}.$$

It follows that,

$$K_p = \frac{p(N_2O_4)}{p(NO_2)^2} = 1.9 \times 10^{-6} \text{ N}^{-1} \text{ m}^2.$$

Question 8

Gaseous hydrogen fluoride is thought to exist as an equilibrium mixture of the monomer HF and the cyclic hexamer $(HF)_6$:

$$6HF(g) \rightleftharpoons (HF)_6(g)$$

(a) Write an expression for the equilibrium constant, K_p, in terms of the partial pressures of the monomer, p(m), and the hexamer, p(h).

(b) A sample of gaseous hydrogen fluoride at equilibrium occupies a 500 cm^3 reaction vessel at one atmosphere pressure at 353 K. Assuming the ideal gas law, calculate the total amount of gaseous molecules present in the sample.

$1 \text{ atm} = 1.01325 \times 10^5 \text{ N m}^{-2}$.

(c) The mass of the sample in part (c) was found to be 0.698 g. Determine the amounts of the hexamer and of the monomer in the sample at equilibrium. [Relative atomic masses: H, 1.0; F, 19.0]

(d) Calculate a value for K_p.

Answer

(a) The expression for the equilibrium constant is

$$K_p = \frac{p(h)}{p(m)^6}$$

(b) To calculate number of moles the ideal gas law is invoked:

$$PV = nRT$$

where P is the pressure, V is the volume, n is the number of moles of gas, T is the absolute temperature and R is the universal gas constant (8.31 J K^{-1} mol^{-1}). Substituting the data into the question it is essential to ensure the compatibility of the units used. One atmosphere corresponds to a pressure of 1.013×10^5 N m^{-2}. Correspondingly the value of 500 cm^{-3} is equivalently 5×10^{-4} m^3. Accordingly

$$1.013 \times 10^5 \times 5 \times 10^{-4} = n \times 8.31 \times 353$$

so that,

$$n = 0.0173 \text{ moles.}$$

(c) The mass of the sample was 0.698 g corresponding to 0.0173 moles. The apparent relative molecular mass of the gas is therefore 0.698/0.0173 (40.44). The relative molecular mass of the monomer is 20 where that of the hexamer is 120. The observed apparent value of the mixture represents a weighted average of the two components:

$$40.44 = 20(1 - x(h)) + 120x(h)$$

where x(h) is the mole fraction of the hexamer present. Solving this linear equation gives x(h) a value of 0.204. Accordingly the number of moles, n(h) of hexamer can be found since

$$x(h) = \frac{\text{no. of moles of h}}{\text{no. of moles of h} + \text{no. of moles of m}}$$
$$= \frac{n(h)}{0.0173} = 0.204.$$

It follows that n(h) has a value of 3.54×10^{-3} moles. The number of moles of the monomer is 0.014 moles.

(d) Knowing the mole fraction of hexamer and monomer present it is possible to evaluate their partial pressures:

$$p(h) = x(h)\, P_{TOT}$$

and

$$p(m) = x(m)\, P_{TOT}$$

where P_{TOT} is the total pressure and x(m) is the mole fraction of monomer. Substituting values gives,

$$p(h) = 0.204 \text{ atm}$$

and

$$p(m) = 0.796 \text{ atm.}$$

The expression for the equilibrium constant,

$$K_p = \frac{p(h)}{p(m)^6} = \frac{0.204}{0.796^6}$$

may be evaluated to give an answer of 0.80 atm^{-5}.

Question 9

(a) The decomposition of phosphorous pentachloride occurs with an enthalpy of reaction in the forward direction of +55 kJ mol^{-1}

$$PCl_5(g) \rightleftharpoons PCl_3(g) + Cl_2(g).$$

Explain what happens when the following occur.
(i) A substance that catalyses the decomposition is added.
(ii) The temperature of the system is increased.
(iii) The pressure of PCl_5 is increased.

(b) PCl_3 is a colourless liquid, whilst PCl_5 consists of white crystals. They can be made as a mixture by heating PCl_5 at 490 K, and subsequently cooling and condensing quickly. Addition of water to the resulting mixture causes hydrolysis of the phosphorous chlorides:

$$PCl_3(\ell) + 3H_2O(\ell) \rightarrow 3HCl(aq) + H_3PO_3(aq)$$
$$PCl_5(s) + H_2O(\ell) \rightarrow 2HCl(aq) + POCl_3(aq)$$

The amount of acid produced can be measured by titration with alkali. In an experiment, one mole of phosphorous pentachloride was heated in a five litre flask. The products were condensed then hydrolysed with water. The resulting solution was made up to 1.00 dm^3. A 25.0 cm^3 aliquot of this solution was titrated with 2.0 M KOH(aq), using bromothymol blue as indicator. 45.0 cm^3 of this solution were required to make the indicator change colour from yellow to blue. Calculate a value for the equilibrium constant for the equilibrium:

A diprotic acid is one that gives up two protons when reacting with base. Sulphuric acid, H_2SO_4, is an example of a diprotic acid.

$$PCl_5(g) \rightleftharpoons PCl_3(g) + Cl_2(g),$$

noting that H_3PO_3 is a diprotic acid.

Answer

(a) (i) Le Chatelier's principle indicates that increasing the pressure will drive the reaction in favour of PCl_5 since reducing the number of moles of gas present would tend to oppose the pressure increase.
(ii) The dissociation is endothermic (as bonds are broken) so that increasing the temperature would shift the reaction in favour of the products.
(iii) A catalyst speeds up a reaction but cannot alter the position of equilibrium. So no change in the equilibrium constant would be expected on introducing a catalyst.

(b) Considering the titration data and given that one mole of OH^- reacts with one mole of H^+ the 1.00 dm^{-3} solution must contain ($2 \times 45 / 25$) moles. The H^+ concentration in the solution is therefore 3.6 M Since 1.00 dm^{-3} of solution is used to make this it follows that one mole of PCl_5 produces 3.6 moles of titratable protons. Examining the stoichiometry of the equilibrium for the hydrolysis of PCl_3 and PCl_5, noting the diprotic character of H_3PO_3, it can be appreciated that the former produces five titratable moles of H^+ for every mole of PCl_3 whereas the latter, PCl_5 produces just two moles. The observed value of 3.6 for the number of protons produced from the hydrolysis reflects a weighted average of the true value:

$$3.6 = 2x + 5(1 - x) \tag{5.7}$$

where x is the fraction of PCl_5 in the gas mixture before condensation and hydrolysis. Solving equation (5.7) gives a value of 0.47 for x.

Returning to the equation describing the reaction of interest and keeping track of the number of moles of the different species before and after the reaction,

	$PCl_5(g)$	$\rightleftharpoons$	$PCl_3(g)$	+	$Cl_2(g)$
Before reaction:	1 mole		0 mole		0 mole
At equilibrium:	0.47 mole		0.53 mole		0.53 mole

To evaluate an equilibrium constant K_p, for the reaction we need to know the partial pressures of the different species. This can be calculated using the ideal gas law:

$$p(PCl_5) = \frac{0.47RT}{V}$$

$$p(PCl_3) = \frac{0.53RT}{V}$$

$$p(Cl_2) = \frac{0.53RT}{V}$$

where V is the volume of the solution: $5 \times 10^{-3}\ m^{-3}$. It follows that,

$$\begin{aligned} K_p &= \frac{p(Cl_2)p(PCl_3)}{p(PCl_5)} \\ &= \frac{0.53^2RT}{0.47\,V} \\ &= 4.9 \times 10^5\ N\ m^{-2} \\ &\approx 4.9\ atm. \end{aligned}$$

Question 10

Methanol is produced commercially by using a mixture of carbon monoxide and hydrogen over a zinc oxide/ copper oxide catalyst

$$CO(g) + 2H_2(g) \rightarrow CH_3OH(g).$$

The standard enthalpy of formation and the absolute entropy for each of the three gases at 298 K and at a pressure of one bar, are given as follows.

1 bar $= 10^5\ N\ m^{-2}$.

	ΔH_f° / kJ mol^{-1}	S° / J K^{-1} mol^{-1}
CO(g)	–111	198
$H_2(g)$	0	131
$CH_3OH(g)$	–201	240

(a) Calculate ΔH°, ΔS°, ΔG° and K_p for the reaction at 298 K.

(b) A commercial reactor is operated at a temperature of 600 K. Calculate the value of K_p at this temperature, assuming $\Delta H_f^\circ(298)$ is independent of temperature.

(c) The gas flowing into the reactor comprises of two moles of H_2 for each mole of CO. If the mole fraction of methanol in the exhaust gas from the reactor is 0.25, what is the total pressure in the reactor?

Assume that equilibrium is established.

Answer

(a) A simple Hess cycle gives a value for ΔH° of -90 kJ mol^{-1}. Similarly,

$$\Delta S^\circ = -220 \text{ J K}^{-1} \text{ mol}^{-1}$$

and

$$\Delta G^\circ = \Delta H^\circ - T\Delta S^\circ$$
$$= -24.4 \text{ kJ mol}^{-1}.$$

A value for the equilibrium constant, K_p, can be found from the expression,

$$\Delta G^\circ = -RT \ln K_p$$

so that,

$$K_p = 1.9 \times 10^4 \text{ atm}^{-2}.$$

(b) To find a value of K_p at 600 K we use the van't Hoff Isochore (see Question 3):

$$\ln K_p = \frac{-\Delta H^\circ}{R} + \text{constant}.$$

It follows that

$$\ln K_p(600) = \ln K_p(298) + \frac{\Delta H^\circ}{R}\left(\frac{1}{298} - \frac{1}{600}\right).$$

Substituting the relevant values gives

$$K_p(600) = 2.2 \times 10^{-4} \text{ atm}^{-2}.$$

The reduced value as compared to 298 K reflects the exothermic nature of the reaction.

(c) Again it is helpful to consider the amounts of different species present before the reaction and when equilibrium is established.

	$CO(g)$	+	$2H_2(g)$	$\rightleftharpoons$	$CH_3OH(g)$
Before reaction	1 mole		2 mole		0 moles
At equilibrium	1 – y mole		2 – 2y mole		y mole

The amount of methanol, y moles, can be ascertained from the fact that the mole fraction of methanol is 0.25 so that

$$0.25 = \frac{\text{no. of moles of methanol}}{\text{no. of moles of hydrogen} + \text{no. of moles of CO} + \text{no. of moles of methanol}}$$
$$= \frac{y}{3 - 2y}$$

so that $y = 0.5$ mole.

From the above it is possible to find the mole fraction x of the different species:

$$x(CH_3OH) = \frac{0.5}{3 - 2 \times 0.5} = 0.25$$

$$x(CO) = \frac{1 - 0.5}{3 - 2 \times 0.5} = 0.25$$

$$x(H_2) = \frac{2 - 2 \times 0.5}{3 - 2 \times 0.5} = 0.50$$

The corresponding partial pressures are

$$p(CH_3OH) = 0.25 \times P_{TOT}$$
$$p(CO) = 0.25 \times P_{TOT}$$

and

$$p(H_2) = 0.50 \times P_{TOT}$$

where P_{TOT} is the total pressure. Since the reactor operates at 600 K

$$K_p = 2.2 \times 10^{-4} = \frac{p(CH_3)}{p(CO)p(H_2)^2}$$
$$= \frac{0.25P_{TOT}}{0.25P_{TOT} \times (0.5P_{TOT})^2}.$$

Solving this equation gives

$$P_{TOT} = 135 \text{ atm}.$$

5.5 Acids and bases

Question 11

(a) The pH of human saliva usually lies in the range pH 6.5–7.6, whilst that for human gastric juices in the range pH 1.0–3.0.
  (i) Explain the term pH.
  (ii) What are the ranges of concentration of $H^+(aq)$ and $OH^-(aq)$ in human saliva and human gastric juices? [The acid dissociation constant for water, $K_w = 1.00 \times 10^{-14}$ mol^2 dm^{-6}.]

(b) (i) How many water molecules are there in a 1.00 dm^3 sample of water at 298 K. How many hydroxonium ions, H_3O^+, are there? [The density of water at 298 K is 0.997 g cm^{-3}.]
  (ii) The forward reaction in the autoionization equilibrium for water is endothermic:

$$2H_2O(\ell) \rightleftharpoons H_3O^+(aq) + OH^-(aq) \qquad \Delta H^\circ > 0.$$

  Explain what happens to the pH of water when the temperature is increased.

Answer

(a) (i) The term pH is related to the concentration of protons in solution:

$$pH = -\log_{10}[H^+(aq)].$$

  (ii) Consider first the proton levels. For human saliva,

$$6.5 < pH < 7.6$$

  so that,

$$10^{-6.5} > [H^+]/M > 10^{-7.6}$$

or

$$3.2 \times 10^{-7} > [H^+]/M > 2.5 \times 10^{-8}.$$

Likewise for gastric juice,

$$1.0 < pH < 3.0$$

so that,

$$10^{-1} > [H^+]/M > 10^{-3}.$$

Turning now to the hydroxide ion concentration, these are related to the above proton concentrations through the equation

$$[OH^-] = K_w/[H^+] = 10^{-14}/[H^+]$$

so that

$$3.2 \times 10^{-8} < [OH^-]/M < 4.0 \times 10^{-7} \quad \text{(saliva)}$$
$$10^{-13} < [OH^-]/M < 10^{-11} \quad \text{(gastric juice)}$$

(b) (i) A 1.00 dm^3 sample of water weighs 997 g. As the relative molecular mass of water is 18 this corresponds to 997/18 (= 55.39) moles. One mole corresponds to an Avogadro number (6.022×10^{23} mol^{-1}) of molecules. So 1.00 dm^3 of water contains

$$55.39 \times 6.022 \times 10^{23} = 3.33 \times 10^{25} \text{ molecules.}$$

In pure water,

$$K_w = [H^+][OH^-] = 10^{-14}$$

and

$$[H^+] = [OH^-]$$

so that

$$[H^+] = 10^{-7} \text{ M.}$$

Accordingly, in 1.0 dm^3 there are 10^{-7} moles of H_3O^+ corresponding to

$$10^{-7} \times 6.022 \times 10^{23} = 6.022 \times 10^{16} \text{ molecules.}$$

(ii) As the dissociation

$$2H_2O(\ell) \rightleftharpoons H_3O^+(aq) + OH^-(aq)$$

is endothermic, increasing the temperature will promote the formation of H^+ and OH^- ions by Le Chatelier's principle. It follows that the pH of the solvent will decrease, since

$$pH = -\log_{10}[H_3O^+].$$

Note that this decrease does not imply a loss of electroneutrality: the increase in H_3O^+ is matched by an identical increase in OH^-.

Question 12

Hydrogen peroxide, H_2O_2, ionises in aqueous solution, and thereby acts as a weak acid:

$$H_2O_2(aq) + H_2O(\ell) \rightleftharpoons H_3O^+(aq) + O_2H^-(aq).$$

The acid dissociation constant, K_a, for hydrogen peroxide is 2.40×10^{-12} mol dm^{-3} at 298 K.

(a) Write down an expression for K_a.

(b) A strong acid was added to an aqueous solution of hydrogen peroxide to give an equilibrium concentration of hydroxonium ions of 0.039 mol dm^{-3}, whilst that of hydrogen peroxide was 0.100 mol dm^{-3}. What is the equilibrium concentration of O_2H^- ions?

Answer

(a) The expression for the acid dissociation constant of hydrogen peroxide is:

$$K_a = \frac{[HO_2^-][H_3O^+]}{[H_2O_2]}.$$

(b) Substituting the values gives,

$$2.40 \times 10^{-12} = \frac{0.039[HO_2^-]}{0.1},$$

so that

$$[HO_2^-] = 6.15 \times 10^{-12} \text{ M}.$$

Question 13

(a) Ethanoic acid ionises as follows:

$$CH_3COOH(aq) + H_2O(\ell) \rightleftharpoons CH_3COO^-(aq) + H_3O^+(aq)$$

The pK_a of ethanoic acid is 4.76 at 298 K.

(i) What is meant by the term pK_a?

(ii) Calculate the pH of a 0.1 M solution of ethanoic acid.

(iii) Calculate the relative amounts of CH_3COOH and CH_3COO^- in a solution of pH = 4.00 at 298 K.

(iv) The electrical conductivity of ethanoic acid in water depends on its concentration in solution. Comment.

(b) (i) Explain why aqueous solutions of aluminium chloride are acidic.

(ii) The value of pK_a for $Al(H_2O)_6^{3+}$ is 4.9. Calculate the pH of a 10^{-3} M solution of a soluble aluminium salt.

Answer

(a) (i) The pK_a of a weak acid, HA, refers to the dissociation,

$$HA(aq) + H_2O(\ell) \rightleftharpoons H_3O^+(aq) + A^-(aq)$$

for which the equilibrium constant,

$$K_a = \frac{[H_3O^+][A^-]}{[HA]}.$$

The pK_a is defined by

$$pK_a = -\log_{10}K_a.$$

(ii) Since the pK_a of ethanoic acid is 4.76, it follows that,

$$K_a = 10^{-4.76} = 1.74 \times 10^{-5}\ M.$$

For the dissociation of a 0.1 M solution of ethanoic acid

$$CH_3COOH(aq) \rightleftharpoons CH_3COO^-(aq) + H^+(aq),$$

the concentration of H^+ and ethanoate ions can be assumed to be the same:

$$[H^+] = [CH_3COO^-]$$

so that,

$$K_a = \frac{[H^+][CH_3COO^-]}{[CH_3COOH]} = \frac{[H^+]^2}{[CH_3COOH]}.$$

Since ethanoic acid is a weak acid, we can first derive an approximate answer by assuming the concentration of ethanoic acid is unchanged at equilibrium from that before the reaction. Accordingly,

$$K_a = 1.74 \times 10^{-5} = \frac{[H^+]^2}{0.1}$$

so that,

$$1.74 \times 10^{-6} = [H^+]^2$$

and hence,

$$[H^+] = 1.32 \times 10^{-3}\ M.$$

The pH may therefore be calculated as

$$\begin{aligned} pH &= -\log_{10}[H^+] \\ &= -\log_{10}(1.32 \times 10^{-3}) \\ &= 2.88 \end{aligned}$$

More rigorously, we must calculate the amount of each species before and after equilibration:

	CH_3COOH	$\rightleftharpoons$	CH_3COO^-	+	H^+
Conc. before reaction / M	0.1		0		0
Conc. after reaction / M	$0.1(1-\alpha)$		0.1α		0.1α

It follows that,

$$K_a = \frac{(0.1\alpha)^2}{0.1(1-\alpha)} = \frac{0.1\alpha^2}{1-\alpha}$$

so that,

$$0.1\alpha^2 + K_a\alpha - K_a = 0$$

or

$$\alpha^2 + 1.74 \times 10^{-4}\alpha - 1.74 \times 10^{-4} = 0$$

This quadratic solves to give

$$\alpha = 0.0133.$$

Accordingly,

$$[H^+] = 0.1\alpha$$
$$= 1.33 \times 10^{-3}\ M,$$

showing the validity of the approximation made above in the approximate answer in respect of ethanoic acid being a weak acid.

(iii) Since

$$K_a = \frac{[H^+][CH_3COO^-]}{[CH_3COOH]},$$

it follows that

$$\log_{10}K_a = \log_{10}[H^+] + \log_{10}\frac{[CH_3COO^-]}{[CH_3COOH]}$$

or,

$$pK_a = pH - \log_{10}\frac{[CH_3COO^-]}{[CH_3COOH]}.$$

This is the Henderson-Hasselbach equation. When the pH is 4.00, and since pK_a is 4.76, it follows that

$$4.76 = 4.00 - \log_{10}\frac{[CH_3COO^-]}{[CH_3COOH]},$$

so that,

$$\frac{[CH_3COO^-]}{[CH_3COOH]} = 0.17.$$

(iv) For the dissociation,

$$K_a = \frac{[H^+][CH_3COO^-]}{[CH_3COOH]}$$

we can again track the concentration before and after equilibration. If the total concentration of ethanoic acid dissolved is C_{TOT}:

	CH_3COOH	$\rightleftharpoons$	CH_3COO^-	+	H^+
Conc. before reaction / M	C_{TOT}		0		0
Conc. after reaction / M	$C_{TOT}(1-\alpha)$		αC_{TOT}		αC_{TOT}

Hence,

$$K_a = \frac{\alpha^2}{1-\alpha} C_{TOT}.$$

It follows that if C_{TOT} is decreased, α must increase to ensure K_a is unchanged. It follows that the fraction of ethanoic acid which is ionised is greater at higher dilution. This is reflected in measurements of conductivity which correlate with the number of ions present.

(b) (i) The acidity results from the dissociation:

$$Al(H_2O)_6^{3+}(aq) \rightleftharpoons Al(H_2O)_5(OH)^{2+}(aq) + H^+(aq).$$

(ii) If the pK_a of $Al(H_2O)_6^{3+}$ is 4.9 then,

$$K_a = 10^{-4.9} = 1.26 \times 10^{-5} = \frac{[H^+][Al(H_2O)_5(OH)^{2+}]}{[Al(H_2O)_6^{3+}]}.$$

To find the pH of a 10^{-3} M solution of aluminium chloride, we again track concentrations before and after equilibration:

	$Al(H_2O)_6{}^{3+}$	$\rightleftharpoons$	$Al(H_2O)_5(OH)^{2+}$	+	H^+
Conc. before reaction / M	10^{-3}		0		0
Conc. after reaction / M	$10^{-3}(1-\alpha)$		$\alpha 10^{-3}$		$\alpha 10^{-3}$

It follows that

$$K_a = 1.26 \times 10^{-5} = \frac{10^{-3}\alpha^2}{1-\alpha}$$

or,

$$10^{-3}\alpha^2 + 1.26 \times 10^{-5}\alpha - 1.26 \times 10^{-5} = 0$$

which solves to give $\alpha = 0.106$, so that

$$[H^+] = 10^{-3} \times 0.106$$
$$= 1.06 \times 10^{-4} \text{ M}.$$

Finally,

$$pH = -\log_{10}[H^+] = 3.97.$$

Question 14

An acid HX ionises in water according to the equation:

$$HX(aq) \rightleftharpoons H^+(aq) + X^-(aq).$$

When differing amounts of HX were dissolved in 100 cm^3 of water, the following concentrations of H^+ ions were measured:

Amount of HX / mol	1.00×10^{-6}	1.00×10^{-4}	1.00×10^{-2}
$[H^+]$ / mol dm^{-3}	7.10×10^{-6}	1.25×10^{-4}	1.29×10^{-3}

(a) Obtain a value for the ionisation constant (dissociation constant) of HX.

(b) 0.01 mol HX is added to 100 cm^3 of 0.2 M NaX(aq) and allowed to come to equilibrium. Estimate the pH of the solution, neglecting any change in volume on the addition.

Answer

(a) For the dissociation constant,

$$K_a = \frac{[H^+][X^-]}{[HX]}.$$

We yet again track concentration before and after equilibration, noting that the moles specified in the data within the question were dissolved in 0.1 dm^3. Initially we consider the first column of data.

	HX	$\rightleftharpoons$	H^+	+	X^-
Conc. before reaction / M	10.00×10^{-6}		0		0
Conc. after reaction / M	2.9×10^{-6}		7.1×10^{-6}		7.1×10^{-6}

It follows that

$$K_a = (7.1 \times 10^{-6})^2/2.9 \times 10^{-6} = 1.74 \times 10^{-5}\ M.$$

Applying exactly analogous proceedures to the other two columns of data, closely similar values of K_a are discovered. It may be concluded that

$$K_a \approx 1.80 \times 10^{-5}\ M.$$

(b) If attention is initially focused solely on the acid dissociation reaction and the speciation considered before and after equilibration whilst noting that 0.01 moles of HX are added to 100 cm^3 of solution,

	HX	$\rightleftharpoons$	H^+	+	X^-
Conc. before reaction / M	10^{-3}		0		0.2
Conc. after reaction / M	$10^{-3}(1-\alpha)$		$10^{-3}\alpha$		$0.2 + 10^{-3}\alpha$

If these values are substituted into the expression for K_a and α is evaluated, it is discovered that $[H^+] \sim 9 \times 10^{-8}$ M, which is less than in pure water! It follows that the analysis must include *both* the acid dissociation constant and the self-ionisation of water:

$$H_2O(\ell) \rightleftharpoons H^+(aq) + OH^-(aq).$$

It is known that

$$K_w = [H^+][OH^-] \tag{5.8}$$

and

$$K_a = \frac{[H^+][X^-]}{[HX]}. \tag{5.9}$$

Also,

$$[X^-] + [HX] = 0.201 \tag{5.10}$$

and, by electroneutrality,

$$[Na^+] + [H^+] = [OH^-] + [X^-] \tag{5.11}$$

From equations (5.9) and (5.10),

$$[X^-] + \left(\frac{[H^+]}{K_a}[X^-]\right) = 0.201$$

so

$$[X^-] = \frac{0.201}{1 + [H^+]/K_a} \tag{5.12}$$

Substituting equations (5.8) and (5.12) into (5.11) together with the initial Na^+ concentration,

$$0.2 + [H^+] = \frac{K_w}{[H^+]} + \frac{0.201}{1 + [H^+]/K_a}$$

Assuming $K_a \ll [H^+]$,

$$[H^+]^2 + 0.2[H^+] - (K_w + 0.201\ K_a) = 0$$

or, since $K_w \ll 0.201\ K_a$

$$[H^+]^2 + 0.2[H^+] - 0.201\ K_a \approx 0$$

This solves to give $[H^+] = 1.8 \times 10^{-5}$ M, which corresponds to pH = 4.7

Question 15

(a) Explain how the pH of a solution might be measured.

(b) An acid-base titration involves the addition of an alkaline solution to a solution of acid (or vice versa). This can allow the concentration of an acidic (or alkaline) solution to be determined.

In an experiment, 50.0 cm^3 of a 0.100 M HCl solution were used to titrate various volumes of 0.100 M NaOH solution at 298 K.

(i) Calculate the pH of the resulting mixture in the flask after the addition of 0, 49.0, 51.0 and 60.0 cm^3 of the NaOH solution, given that $K_w = 1.00 \times 10^{-14}$ mol^2 dm^{-6}.

(ii) Draw a graph using pH as the vertical axis and the volume of NaOH solution as the horizontal axis. Sketch the titration curve.

(iii) Predict qualitatively the form of the titration curve when 0.10 M ethanoic acid is titrated with 0.10 M NaOH.

(iv) How can a titration curve be used to estimate the acid dissociation constant, K_a, of a weak acid such as ethanoic acid?

(c) Maleic acid is a *diprotic* acid with ionisation equilibria having the values $pK_{a1} = 1.9$ and $pK_{a2} = 6.2$ in water at 298 K. Fig. 5.6 shows how the pH of a 25 cm^3 solution of 0.1 M maleic acid varies as the acid is titrated with 0.1 M NaOH. Discuss the features shown on the curve.

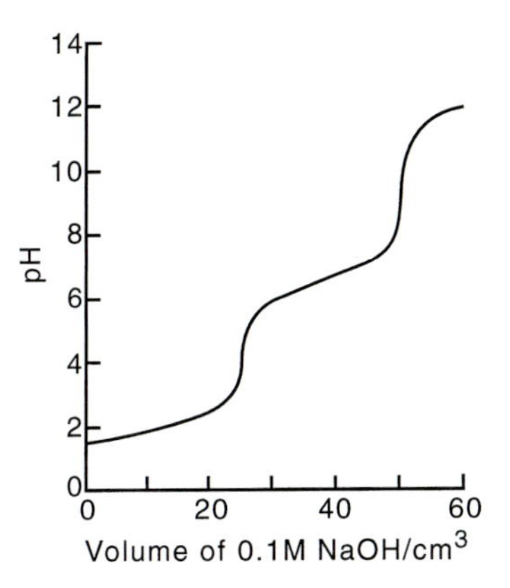

Fig. 5.6 pH titration curve for maleic acid.

(d) The pH of fluids in living creatures needs to be maintained within a constant value. Such control can be exercised by buffer solutions.

Human blood plasma contains a bicarbonate buffer, which ensures that blood pH is in the range pH 7.3–7.5. This buffer can be represented by the equilibrium:

$$H^+(aq) + HCO_3^-(aq) \rightleftharpoons H_2CO_3(aq) \rightleftharpoons H_2O(\ell) + CO_2(g)$$

(i) Explain what is meant by the term buffer?

(ii) Explain how the acidity of human blood can be affected by deep or shallow breathing.

Blood acid levels are effectively maintained by the renal system using several buffers which include bicarbonate, phosphate and ammonium ions. In the event of too low pH (acidosis), both $H_2PO_4^{2-}$ and ammonium ions are excreted in urine, until the acid base balance is restored. In the opposite case, alkalosis, the reverse occurs, but it is mainly HPO_4^- that is excreted.

Answer

(a) The pH of a solution can be measured in several ways. The simplest is to employ an indicator which changes colour above or below a known pH value. More accurately, a glass electrode is commonly used as the key part of a pH meter. This contains a thin glass membrane in which there are Na^+ cations which can exchange with protons in a solution bathing the membrane:

$$H^+(aq) + Na^+(glass) \rightleftharpoons H^+(glass) + Na^+(aq).$$

Depending on the extent of proton uptake and sodium loss, a charge develops at the glass-solution interface which is measured as an electrode potential by the pH meter. By calibrating the pH meter using suitable solutions of known composition, reliable measurements to within 0.01 of a pH unit are routinely accessible.

(b) (i) Before any addition of NaOH to a 0.1 M solution of HCl, the pH of the latter is:

$$\begin{aligned} pH &= -\log_{10}[H^+] \\ &= -\log_{10}10^{-1} \\ &= 1 \end{aligned}$$

When 49 cm^3 of base has been added there remains, in effect just 1 cm^3 of unreacted H^+, now in a volume of 99 cm^3. The concentration of protons is therefore

$$[H^+]_{49} = \frac{1}{99} \times 10^{-1} = 1.01 \times 10^{-3}\ M,$$

corresponding to a pH of 3.0. Likewise, the addition of 51 cm^3 of base leaves in effect 1 cm^3 of unreacted OH^- of a concentration

$$[OH^-] = \frac{1}{101} \times 10^{-1} = 0.99 \times 10^{-3}\ M.$$

It follows that

$$[H^+]_{51} = \frac{K_w}{0.99 \times 10^{-3}} = 10^{-11}\ M.$$

The pH is therefore 11. Last, the addition of 60 cm^3 produces effectively 10 cm^3 of 0.1 M OH^- diluted to 110 cm^3 corresponding to

$$[OH^-] = \frac{1}{110} \times 10^{-1} = 0.91 \times 10^{-2}\ M.$$

The associated proton concentration is:

$$[H^+]_{60} = \frac{K_w}{0.91 \times 10^{-2}} = 1.1 \times 10^{-12}\ M.$$

This corresponds to a pH of approximately 12.

(ii) The titration curve is shown in Fig. 5.7.

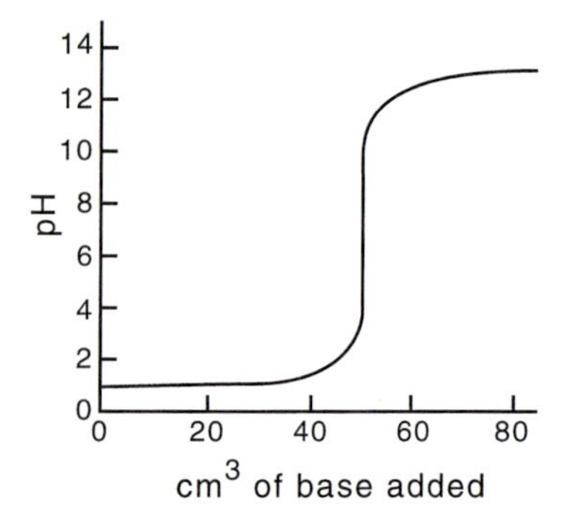

Fig. 5.7 Titration curve for strong acid-strong base.

(iii) The form of the weak acid-strong base titration curve is illustrated in Fig. 5.8. This deviates from the strong acid-strong base case described in part (ii) in the region where pH < 7. This is because the equilibrium

$$CH_3COOH(aq) \rightleftharpoons CH_3COO^-(aq) + H^+(aq)$$

is established which serves to "buffer" the solution at pH values not too far from the pK_a of the weak acid. The addition of base therefore serves to remove protons from undissociated ethanoic acid rather than "mop up" free protons from the solution.

(iv) Returning to the Henderson-Hasselbach equation (Question 11)

$$pH = pK_a + \log_{10}\frac{[CH_3COO^-]}{[CH_3COOH]},$$

it can be seen that pH = pK_a when $[CH_3COO^-] = [CH_3COOH]$. This corresponds to the midpoint of Fig. 5.8.

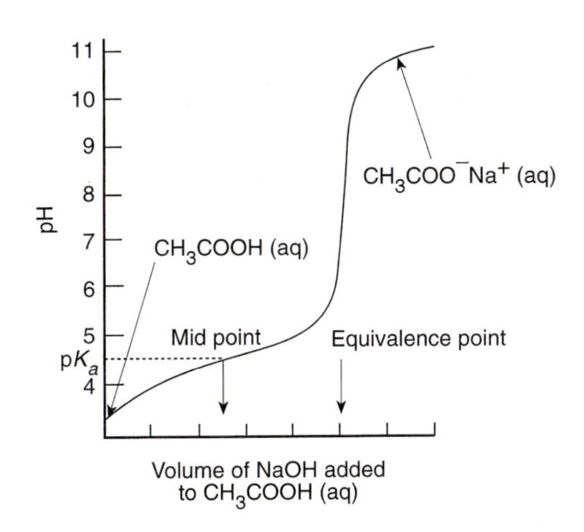

Fig. 5.8 Titration curve for weak acid-strong base showing how K_a can be estimated for ethanoic acid.

(c) Fig. 5.6 shows two end points at volumes of added sodium hydroxide of approximately 25 and 50 cm^3 respectively. These correspond to the sequential removal of the two protons from maleic acid, H_2M:

$$H_2M(aq) \rightleftharpoons HM^-(aq) + H^+(aq) \qquad pK_{a1} = 1.8$$
$$HM^-(aq) \rightleftharpoons M^{2-}(aq) + H^+(aq) \qquad pK_{a2} = 6.1$$

The two pK_a values correspond to the two "mid-points" at around 12.5 and 37.5 cm^3 respectively as in part (b) above.

(d) (i) A buffer is a conjugate acid/base system whose equilibrium can shift to absorb of release protons, thus keeping the pH approximately constant. Any buffer contains significant concentrations of a weak acid and its conjugate base, usually from an added salt.

(ii) Shallow breathing would build up CO_2 in the lungs so pushing the carbonate/bicarbonate equilibrium in a direction to increase the proton concentration. Deep breathing would vent CO_2 from the lungs and have the opposite effect.

Question 16

(a) What is meant by a Lowry-Brønsted acid and a Lowry-Brønsted base?
(b) Define the quantity pK_b applied to a base.
(c) The pK_a of $C_2H_5CO_2H$ is 4.9 and the pK_b of $C_6H_5NH_2$ (aniline) is 9.4. Use these data to predict the position of the following equilibrium at 298 K.

$$C_2H_5CO_2H(aq) + C_6H_5NH_2(aq) \rightleftharpoons C_2H_5CO_2^-(aq) + C_6H_5NH_3^+(aq)$$

Answer

(a) The Lowry-Brønsted theory of acids and bases defines:

- an acid as a proton donor—it is a source of H^+ ions, and
- a base as a proton acceptor—it receives H^+ ions.

(b) For a base A^-,

$$A^-(aq) + H_2O(\ell) \rightleftharpoons HA(aq) + OH^-(aq)$$

the base dissociation (ionisation) constant is defined as

$$K_b = \frac{[HA][OH^-]}{[A^-]}.$$

The term,

$$pK_b = -\log_{10}K_b.$$

(c) The pK_a of propanoic acid given implies that

$$10^{-4.9} = 1.3 \times 10^{-5} = \frac{[C_2H_5COO^-][H_3O^+]}{[C_2H_5COOH]},$$

whereas the pK_b of aniline suggests that

$$10^{-9.4} = 4.0 \times 10^{-10} = \frac{[C_6H_5NH_3^+][OH^-]}{[C_6H_5NH_2]}.$$

Considering the equilibrium

$$C_2H_5CO_2H(aq) + C_6H_5NH_2(aq) \rightleftharpoons C_2H_5CO_2^-(aq) + C_6H_5NH_3^+(aq),$$

the equilibrium constant,

$$K = \frac{[C_6H_5NH_3^+][C_2H_5CO_2^-]}{[C_2H_5CO_2H][C_6H_5NH_2]}$$
$$= \frac{4.0 \times 10^{-10}}{[OH^-]} \times \frac{1.3 \times 10^{-5}}{[H_3O^+]}.$$

But $[OH^-][H^+] = K_w$, so that

$$K = \frac{4.0 \times 10^{-10} \times 1.3 \times 10^{-5}}{10^{-14}} = 0.52.$$

Question 17: Diprotic acids

(a) *cis*-Butenedioic (maleic) acid $\{HO_2CC(H)=C(H)CO_2H\}$, H_2A, is a diprotic acid.
It has two pK_a values that relate to the following equilibria.

$$H_2A(aq) + H_2O(\ell) \rightleftharpoons HA^-(aq) + H_3O^+(aq) \qquad pK_{a1} = 1.8$$
$$HA^-(aq) + H_2O(\ell) \rightleftharpoons A^{2-}(aq) + H_3O^+(aq) \qquad pK_{a2} = 6.1$$

Consider a solution formed by dissolving 10^{-3} moles of *cis*-butenedioic acid in one cubic decimetre of water. If the pH of the solution is changed, for example by adding mineral acid or base, the relative amounts of H_2A, HA^- and A^{2-} will change. Noting that

$$[H_2A] + [HA^-] + [A^{2-}] = 10^{-3} \text{ mol dm}^{-3},$$

sketch graphs which show how concentration of each of the species H_2A, HA^- and A^{2-} varies with the pH of the solution over the range $0 < pH < 10$.

(b) *trans*-Butenedioic (fumaric) acid has $pK_{a1} = 3.0$ and $pK_{a2} = 4.4$ at 298 K. Why do these values differ from those of *cis*-butenedioic acid?

Answer

(a) By definition of the two pK_a quantities,

$$10^{-1.8} = \frac{[HA^-][H_3O^+]}{[H_2A]} = K_{a1}$$

and

$$10^{-6.1} = \frac{[A^{2-}][H_3O^+]}{[HA^-]} = K_{a2}.$$

Now, it is given that

$$[H_2A] + [HA^-] + [A^{2-}] = 10^{-3}\ M.$$

Substituting for $[H_2A]$,

$$\frac{[HA^-][H_3O^+]}{K_{a1}} + [HA^-] + [A^{2-}] = 10^{-3}.$$

Eliminating $[HA^-]$,

$$\left(\frac{[H_3O^+]}{K_{a1}} + 1\right)\frac{[A^{2-}][H_3O^+]}{K_{a2}} + [A^{2-}] = 10^{-3}.$$

It follows that

$$[A^{2-}] = \frac{10^{-3}}{\left(\frac{[H_3O^+]}{K_{a1}} + 1\right)\left(\frac{[H_3O^+]}{K_{a2}} + 1\right)}.$$

This allows $[A^{2-}]$ to be found as a function of H^+ or pH. $[HA^-]$ can then be found using the expression for K_{a2} above. Finally, the equation for K_{a1} allows the pH dependence of $[H_2A]$ to be determined. The resulting speciation for maleic acid is given in Fig. 5.9.

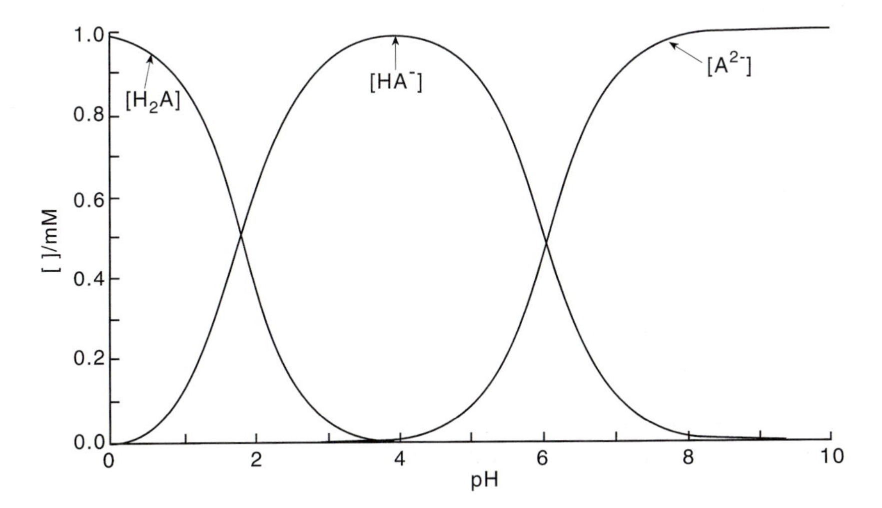

Fig. 5.9 Speciation curve for maleic acid.

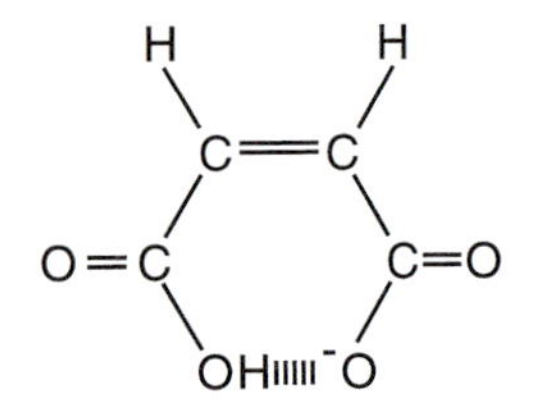

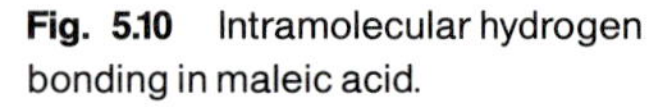

Fig. 5.10 Intramolecular hydrogen bonding in maleic acid.

(b) The pK_a data for *cis*- and *trans*-butenedioic acids implies that the first proton is more readily lost from the cis isomer than the trans isomer, but that the second proton is harder to remove from the cis species. This implies that the mono-protonated anion is stabilised in the cis case as compared to the trans case. This is due to intramolecular hydrogen bonding as shown in Fig. 5.10.

5.6 Solubility and solution equilibria

Question 18

(a) Ag_2CrO_4 is slightly soluble in water as $Ag^+(aq)$ and $CrO_4^{2-}(aq)$ ions:

$$Ag_2CrO_4(s) \rightleftharpoons 2Ag^+(aq) + CrO_4^{2-}(aq).$$

The solubility product, $K_{sp} = [Ag^+]^2[CrO_4^{2-}] = 1.12 \times 10^{-12}$ mol^3 dm^{-9}. For a saturated solution of Ag_2CrO_4 in water calculate:

(i) the molar concentration of silver ions, $[Ag^+]$, in the solution,

(ii) the molar concentration of chromate ions, $[CrO_4^{2-}]$, in the solution, and

(iii) the solubility of Ag_2CrO_4 in g dm^{-3}.

[Relative atomic masses: Ag, 108.0; Cr, 52.0; O, 16.0]

(b) The value of the solubility product at 298 K for $Cu_3(PO_4)_2$ is 1.4×10^{-37} mol^5 dm^{-15}. Calculate its solubility (in mol dm^{-3}). What is the solubility of $Cu_3(PO_4)_2$ in 0.1 mol dm^{-3} $CuSO_4$?

Answer

(a) (i) Given the dissolution reaction,

$$Ag_2CrO_4(s) \rightleftharpoons 2Ag^+(aq) + CrO_4^{2-}(aq)$$

it follows that in the resulting solution,

$$[Ag^+] = 2[CrO_4^{2-}]$$

Now,

$$K_{sp} = [Ag^+]^2[CrO_4^{2-}] = \tfrac{1}{2}[Ag^+]^3$$

so that the molar concentration of Ag^+ is

$$(2K_{sp})^{\frac{1}{3}} = 1.3 \times 10^{-4}\ \text{M}.$$

(ii) Alternatively,

$$K_{sp} = 4[CrO_4^{2-}]^3$$

so that the molar concentration of CrO_4^{2-} is

$$(K_{sp}/4)^{\frac{1}{3}} = 6.54 \times 10^{-5}\ \text{M},$$

noting again that $[Ag^+] = 2[CrO_4^{2-}]$.

(iii) The relative molecular mass of Ag_2CrO_4 is

$$2 \times 108 + 52 + 4 \times 16 = 332.$$

If the concentration of dissolved CrO_4^{2-} is 6.54×10^{-5} M, then the solubility is

$$332 \times 6.54 \times 10^{-5} = 0.022 \text{ g dm}^{-3}.$$

(b) For $Cu_3(PO_4)_2$, the dissociation reaction is:

$$Cu_3(PO_4)_2(s) \rightleftharpoons 3Cu^{2+}(aq) + 2PO_4^{2-}(aq).$$

so that

$$2[Cu^{2+}] = 3[PO_4^{2-}]$$

and,

$$K_{sp} = [Cu^{2+}]^3[PO_4^{2-}]^2 = 1.4 \times 10^{-37} \text{ M}^5.$$

It follows that at equilibrium,

$$K_{sp} = \frac{4}{9}[Cu^{2+}]^5$$

so that

$$[Cu^{2+}] = (9K_{sp}/4)^{\frac{1}{5}}$$
$$= 5 \times 10^{-8} \text{ M}.$$

Accordingly, the solubility of the salt in mol dm^{-3} is one third of this value: 1.7×10^{-8} M.

Last, we consider the solubility of $Cu_3(PO_4)_2$ in a 0.1 mol dm^{-3} $CuSO_4$ solution and track the concentration before and after dissolution:

	$Cu_3(PO_4)_2$	$\rightleftharpoons$	$3Cu^{2+}$	+	$2PO_4^{2-}$
Conc. before reaction / M	0		0.1		0
Conc. after reaction / M	0		0.1 + 3x		2x

where x is the moles per dm^3 of the salt which dissolves. Accordingly,

$$K_{sp} = (0.1 + 3x)^3(2x)^2.$$

This is a quintic equation which is not trivially amenable to immediate solution. However, simplification occurs on noting that the insolubility of the salt suggests

$$3x \ll 0.1$$

so that

$$K_{sp} = (0.1)^3(2x)^2$$

and

$$x = 5.9 \times 10^{-18} \text{ M}.$$

The solubility is therefore much reduced in comparison with that in pure water. This is the common ion effect.

Question 19

Chromate (VI) ions, CrO_4^{2-}, are yellow and in acidic solution exist in equilibrium with dichromate (VI) ions, $Cr_2O_7^{2-}$, which are orange:

$$2CrO_4^{2-}(aq) + 2H^+(aq) \rightleftharpoons Cr_2O_7^{2-}(aq) + H_2O(\ell).$$

(a) (i) Write an expression for the equilibrium constant, K.
(ii) The forward reaction has a standard enthalpy of reaction, $\Delta H^\circ = -14$ kJ mol^{-1} and standard entropy of reaction, $\Delta S^\circ = +231$ J K^{-1} mol^{-1} at 298 K.
(A) Comment on the sign of ΔS°.
(B) Determine ΔG° and K for the above reaction at 298 K.
(C) Calculate the pH of a 1.0 M solution of sodium dichromate (VI).

(b) (i) $BaCrO_4$ is a sparingly soluble salt, but $BaCr_2O_7$ dissolves in water at 298 K. Derive an expression for the equilibrium pH of sodium dichromate (VI) in the presence of barium chromate (VI).
(ii) Use the following data, taken at a constant temperature, to estimate graphically the solubility (in mol dm^{-3}) of barium chromate (VI) at this temperature.

$[Cr_2O_7^{2-}]$ / mol dm^{-3}	0.15	0.40	0.60	0.91
$\dfrac{[Ba^{2+}]}{[H^+]}$	0.526	0.323	0.260	0.211

Answer

(a) (i) The equilibrium constant is:

$$K = \frac{[Cr_2O_7^{2-}]}{[CrO_4^{2-}]^2[H^+]^2}.$$

(ii) (A) The sign of ΔS° is positive reflecting that in the reaction, two doubly charged ions and two protons, all of which are strongly solvated, react to form a single doubly charged ion. Water is therefore released from the solvation shells of the ions in the reaction and the extra entropy associated with this free water causes the large and positive value of ΔS°.

(B) Knowing ΔH° and ΔS°, it follows that

$$\begin{aligned}\Delta G^\circ &= \Delta H^\circ - T\Delta S^\circ \\ &= -14 - 298 \times 231 \times 10^{-3} \\ &= -82.8 \text{ kJ mol}^{-1}.\end{aligned}$$

Since,

$$\Delta G^\circ = -RT \ln K$$

then,

$$K = \exp\left(\frac{82.8 \times 10^3}{8.31 \times 298}\right) = 3.4 \times 10^{14} \text{ M}^{-3}.$$

(C) To find the pH of the solution, the concentrations of the ions must be tracked before and after equilibration.

	$2CrO_4^{2-}$	+	$2H^+$	$\rightleftharpoons$	$Cr_2O_7^{2-}$
Conc. before reaction / M	0		0		1.0
Conc. after reaction / M	2x		2x		1.0 − x

It follows that

$$K = \frac{(1.0 - x)}{(2x)^4}.$$

Assuming that $x \ll 1$, it can be seen that

$$x = \frac{1}{2K^{\frac{1}{4}}} = 1.2 \times 10^{-4}\ M.$$

Therefore,

$$[H^+] = 2.3 \times 10^{-4}\ M$$

and

$$pH = -\log_{10}[H^+] = 3.6.$$

(b) (i) It was established above that

$$K = \frac{[Cr_2O_7^{2-}]}{[CrO_4^{2-}]^2[H^+]^2}.$$

Also,

$$K_{sp} = [Ba^{2+}][CrO_4^{2-}]$$

where K_{sp} is the solubility product of the sparingly soluble $BaCrO_4$. Substituting for $[CrO_4^{2-}]$,

$$K = \frac{[Cr_2O_7^{2-}]}{K_{sp}^2}\frac{[Ba^{2+}]^2}{[H^+]^2} \tag{5.13}$$

or

$$[H^+]^2 = \frac{[Cr_2O_7^{2-}][Ba^{2+}]^2}{K_{sp}^2 K}$$

so that

$$pH = -\frac{1}{2}\log_{10}\frac{[Cr_2O_7^{2-}][Ba^{2+}]^2}{K_{sp}^2 K}.$$

(ii) Rearranging equation (5.13)

$$\frac{[Ba^{2+}]^2}{[H^+]^2} = \frac{K_{sp}^2 K}{[Cr_2O_7^{2-}]}.$$

This equation encourages us to plot a graph of $\frac{[Ba^{2+}]^2}{[H^+]^2}$ against $[Cr_2O_7^{2-}]^{-1}$. This plot is given in Fig. 5.11; the straight through the origin is consistent with the above arguments, and from the slope of the best fit line,

$$\text{slope} = K_{sp}^2 K = 0.0417\ M^{-1}.$$

Using the calculated value of K,

$$K_{sp} = 1.1 \times 10^{-8}\ M^2.$$

5.7 Redox chemistry

Question 20

(a) (i) Define the term standard electrode potential as applied to the Fe^{3+}/Fe^{2+} redox couple.

(ii) Describe the construction and function of a standard hydrogen electrode.

(b) Some standard electrode potentials are given below:

Half-Reaction	E° / V
$\frac{1}{2}Br_2(aq) + e^- \rightarrow Br^-(aq)$	+1.09
$Fe^{3+}(aq) + e^- \rightarrow Fe^{2+}(aq)$	+0.77
$\frac{1}{2}Cu^{2+}(aq) + e^- \rightarrow \frac{1}{2}Cu(s)$	+0.34
$\frac{1}{2}Sn^{4+}(aq) + e^- \rightarrow \frac{1}{2}Sn^{2+}(aq)$	+0.15
$H^+(aq) + e^- \rightarrow \frac{1}{2}H_2(g)$	0

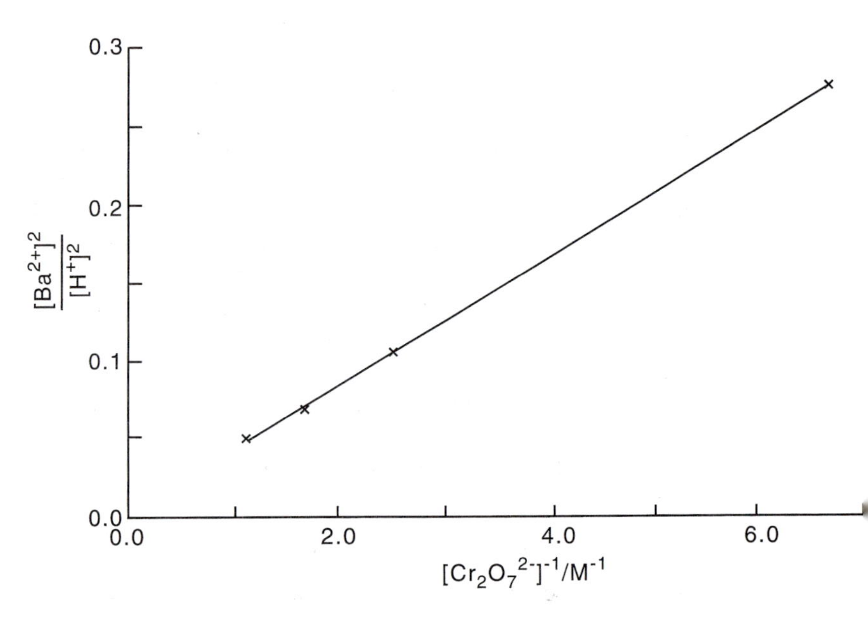

Fig. 5.11 Plot to determine K_{sp} for $BaCrO_4$.

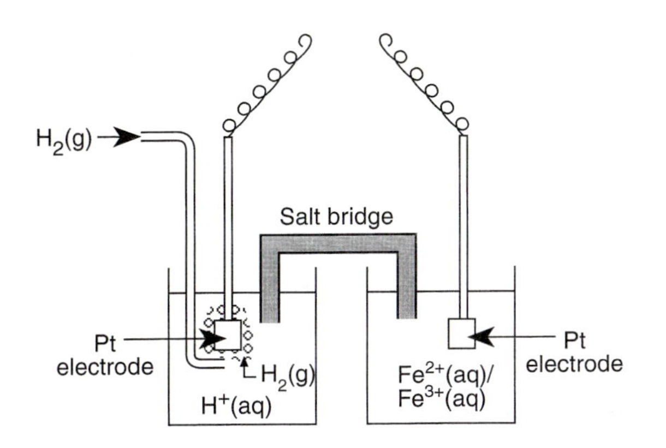

Fig. 5.12 The setup to measure the standard electrode potential of the Fe^{3+}/Fe^{2+} couple, relative to the standard hydrogen electrode.

Notice that the two half-cells in Fig. 5.12 are connected by a salt bridge. This provides an electrical connection betweeen the two half-cells, and is usually, though not invariably, made using concentrated aqueous potassium chloride.

Use these values to predict whether the following equilibria lie in favour of the reactants or the products.

(i) $Sn^{2+}(aq) + Br_2(aq) \rightleftharpoons Sn^{4+}(aq) + 2Br^-(aq)$
(ii) $2Fe^{2+}(aq) + Br_2(aq) \rightleftharpoons 2Fe^{3+}(aq) + 2Br^-(aq)$
(iii) $Sn^{4+}(aq) + 2Fe^{2+}(aq) \rightleftharpoons Sn^{2+}(aq) + 2Fe^{3+}(aq)$
(iv) $Cu(s) + 2H^+(aq) \rightleftharpoons Cu^{2+}(aq) + H_2(g)$

Comment on the observation that no copper is produced when hydrogen gas is bubbled through a copper sulphate solution.

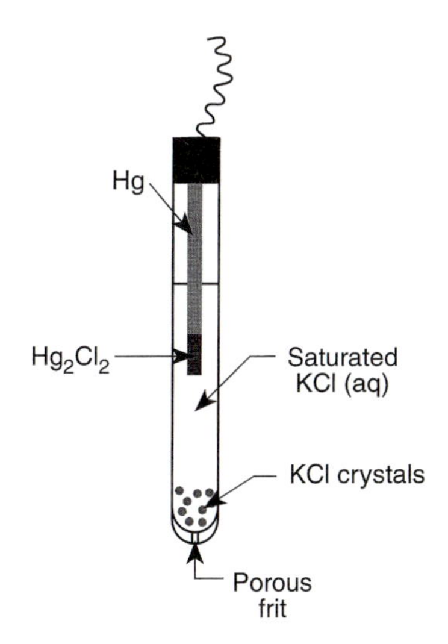

Fig. 5.13 The calomel electrode.

Answer

(a) (i) The standard electrode potential of the Fe^{3+}/Fe^{2+} redox reaction is the potential difference measured between the two electrodes of the cell shown in Fig. 5.12 when all the concentrations of the chemical species shown (H^+, H_2, Fe^{2+}, Fe^{3+}) are unity.

(ii) A standard hydrogen electrode comprises of a platinum electrode coated in platinum black, dipping into a solution of molar hydrochloric acid and over which hydrogen gas is bubbled at a pressure of 1 atm. The device is used at 298 K. The purpose of the standard hydrogen electrode is to act as a reference electrode against which other half-cells can be studied and, the potential difference between the studied couple and the standard hydrogen electrode measured.

Platinum black is finely divided platinum.

Hydrogen gas is inconvenient for routine laboratory use, so the standard hydrogen electrode is commonly replaced by a calomel electrode which has a known potential of 0.24 V relative to the standard hydrogen electrode. A calomel electrode is shown in Fig. 5.13.

(b) The more positive a value of a standard electrode potential the stronger oxidising agent is the corresponding redox couple in question.

(i) The standard electrode potential $E^\circ(Br_2/Br^-)$ is more positive than $E^\circ(Sn^{4+}/Sn^{2+})$ so that Br_2 will oxidise Sn^{2+} to Sn^{4+}, itself being reduced to Br^-. The reaction shown lies in favour of the products.

(ii) In this case

$$E^\circ(Br_2/Br^-) > E^\circ(Fe^{3+}/Fe^{2+})$$

so Br_2 will oxidise Fe^{2+} to Fe^{3+}. The reaction lies in favour of the products.

(iii) In this case

$$E^\circ(Sn^{4+}/Sn^{2+}) < E^\circ(Fe^{3+}/Fe^{2+})$$

so Sn^{4+} will not oxidise Fe^{2+} to Fe^{3+}. The reaction lies in favour of the reactants.

(iv) For this reaction

$$E^\circ(H^+/H_2) < E^\circ(Cu^{2+}/Cu)$$

so that copper metal is not oxidised by H^+ to Cu^{2+}. The equilibrium lies in favour of the reactants.
The last prediction suggests that the reaction

$$\tfrac{1}{2}H_2(g) + \tfrac{1}{2}Cu^{2+}(aq) \rightarrow \tfrac{1}{2}Cu(s) + H^+(aq)$$

is viable. The fact that no reaction is observed suggests that the kinetics of the reaction are very slow.

Question 21

In the cell diagram, the symbol | represents a boundary between two phases (say between the solid electrode and the ionic solution), whilst the symbol || represents a salt bridge.

Consider the following electrochemical cell:

$$Cu(s) \mid Cu^{2+}(aq, 1.00 \text{ mol dm}^{-3}) \parallel Zn^{2+}(aq, 1.00 \text{ mol dm}^{-3}) \mid Zn(s)$$

in the light of the following standard electrode potentials.

Half-Reaction	E° / V
$\tfrac{1}{2}Cu^{2+}(aq) + e^- \rightarrow \tfrac{1}{2}Cu(s)$	+0.34
$\tfrac{1}{2}Zn^{2+}(aq) + e^- \rightarrow \tfrac{1}{2}Zn(s)$	−0.76

(a) Deduce the potential of the cell, making clear which electrode is positive and which is negative.
(b) Predict qualitatively how the cell potential would change if the concentration of Zn^{2+} were decreased.
(c) Predict qualitatively how the cell potential would change if the concentration of Cu^{2+} were increased.
(d) Explain what would happen if a wire was connected between the two electrodes, short-circuiting the cell.
(e) Predict where the position of the following equilibrium lies at 298 K.

$$Cu^{2+}(aq) + Zn(s) \rightleftharpoons Cu(s) + Zn^{2+}(aq)$$

Answer

(a) The pertinent data are

$$E^\circ(Zn^{2+}/Zn) = -0.76 \text{ V}$$

and

$$E^\circ(Cu^{2+}/Cu) = +0.34 \text{ V}.$$

The potential difference of the cell, measured as that of the potential of the zinc electrode relative to that of the copper electrode, is therefore

$$E^\circ(Zn^{2+}/Zn) - E^\circ(Cu^{2+}/Cu) = -1.10 \text{ V}.$$

The zinc electrode is negative and the copper positive. The potential of the two electrodes reflect the following equilibria which are established at the respective electrode-solution interfaces:

$$\tfrac{1}{2}Zn^{2+}(aq) + e^- \rightleftharpoons \tfrac{1}{2}Zn(s)$$

and

$$\tfrac{1}{2}Cu^{2+}(aq) + e^- \rightleftharpoons \tfrac{1}{2}Cu(s).$$

Therefore the former equilibrium lies more to the left than does the latter since the zinc electrode is more negative than the copper electrode.

(b) If the concentration of Zn^{2+} was decreased the equilibrium,

$$\tfrac{1}{2}Zn^{2+}(aq) + e^- \rightleftharpoons \tfrac{1}{2}Zn(s)$$

will shift to the left. This would increase the negative charge (e^-) on the zinc electrode so that the cell potential difference would become even more negative than −1.10 V.

(c) If the concentration of the Cu^{2+} was increased the equilibrium,

$$\tfrac{1}{2}Cu^{2+}(aq) + e^- \rightleftharpoons \tfrac{1}{2}Cu(s)$$

would shift to the right hand side, removing negative charge (e^-) from the copper electrode and so making the copper tend to become relatively more positive. Accordingly the cell potential difference measured as the zinc potential relative to the copper potential would again become more negative than −1.10 V.

(d) If the cell were short circuited electrons would flow from the negative (Zn) towards the positive one (Cu). This would lead to the dissolution of the zinc from the former electrode:

$$\tfrac{1}{2}Zn(s) \rightleftharpoons \tfrac{1}{2}Zn^{2+}(aq) + e^-$$

and the deposition of copper at the latter electrode

$$\tfrac{1}{2}Cu^{2+}(aq) + e^- \rightleftharpoons \tfrac{1}{2}Cu(s).$$

(e) Examining the relative potentials of the two redox couples Zn^{2+} is clearly a less oxidising agent than Cu^{2+}. Therefore Cu^{2+} can oxidise Zn to Zn^{2+}:

$$Cu^{2+}(aq) + Zn(s) \rightleftharpoons Cu(s) + Zn^{2+}(aq)$$

where the equilibrium lies in favour of the products.

Question 22

The Nernst equation for electrode potentials but non-unit concentrations is:

$$E = E^\circ + \frac{RT}{nF}\ln[M^{n+}] \text{ volts},$$

where R is the universal gas constant (8.31 J K^{-1} mol^{-1}),
T is the absolute temperature,
n is the number of moles of electrons involved in the M^{n+}/M redox couple, and
F is the Faraday constant (96 485 C mol^{-1}), and represents the electrical charge per mole of electrons.

The mathematically inclined reader will find a derivation of the Nernst Equation in chapter one, R. G. Compton and G. H. W. Sanders, "Electrode Potentials", (OCP 41).

Find the potentials of the following cells at 298 K:

$$Cu(s)\,|\,Cu(NO_3)_2(aq,\ 1.0\ mol\ dm^{-3})\,||\,AgNO_3(aq,\ 1.0\ mol\ dm^{-3})\,|\,Ag(s)$$
$$Cu(s)\,|\,Cu_3(PO_4)_2(aq\ sat)\,||\,Ag_2CrO_4(aq\ sat)\,|\,Ag(s)$$

where (aq sat) indicates aqueous saturated solution.

The standard electrode potentials are:

$$Ag^+(aq) + e^- \rightleftharpoons Ag(s) \qquad E^\circ = +0.80\ V$$
$$\tfrac{1}{2}Cu^{2+}(aq) + e^- \rightleftharpoons \tfrac{1}{2}Cu(s) \qquad E^\circ = +0.34\ V.$$

The solubility products are:

$$Ag_2CrO_4{:}\quad 1.12 \times 10^{-12}\ mol^3\ dm^{-9}$$
$$Cu_3(PO_4)_2{:}\quad 1.4 \times 10^{-37}\ mol^5\ dm^{-15}.$$

Answer

For the cell

$$Cu(s)\,|\,Cu(NO_3)_2(aq,\ 1.0\ mol\ dm^{-3})\,||\,AgNO_3(aq,\ 1.0\ mol\ dm^{-3})\,|\,Ag(s),$$

the potential of the silver electrode relative to the copper electrode is

$$E^\circ(Ag^+/Ag) - E^\circ(Cu^{2+}/Cu) = 0.46\ V$$

Turning to the second cell,

$$Cu(s)\,|\,Cu_3(PO_4)_2(aq\ sat)\,||\,Ag_2CrO_4(aq\ sat)\,|\,Ag(s)$$

it is necessary to evaluate the concentrations of Ag^+ and Cu^{2+} in the saturated solutions of the salts Ag_2CrO_4 and $Cu(PO_4)_2$. These were calculated in Question 18

$$[Ag^+] = 1.3 \times 10^{-4}\ M$$
$$[Cu^{2+}] = 5 \times 10^{-8}\ M$$

Using the Nernst Equation

$$E = E^\circ + \frac{RT}{nF}\ln[M^{n+}]$$

for silver

$$\begin{aligned} E(Ag^+/Ag) &= E^\circ(Ag^+/Ag) + \frac{RT}{nF}\ln[Ag^+] \\ &= 0.8 + 0.0257\ \ln(1.3 \times 10^{-4}) \\ &= 0.57\ V. \end{aligned}$$

Likewise for copper

$$\begin{aligned} E(Cu^{2+}/Cu) &= E^\circ(Cu^{2+}/Cu) + \frac{RT}{nF}\ln[Cu^{2+}] \\ &= 0.34 + (0.0257/2)\ \ln(5 \times 10^{-8}) \\ &= 0.12\ V \end{aligned}$$

It follows that the cell potential will be given by

$$0.57 - 0.12 = 0.45\ V$$

Question 24

(a) The standard potential of the Cu^+/Cu couple is +0.52 V, whilst that of the Cu^{2+}/Cu couple is +0.34 V at 298 K. Calculate the equilibrium constant for the disproportionation of Cu^+ in water.

(b) Generalise the case of part (a) to consider any metal ion, $M^+(aq)$, which disproportionates in water to give the metal, M(s), and its doubly charged cation, $M^{2+}(aq)$. What can be deduced about the standard potentials of the $M^+(aq) \mid M(s)$ and $M^{2+}(aq) \mid M(s)$ couples?

Answer

(a) The standard electrode potentials of the Cu^{2+}/Cu couple,

$$E^\circ(Cu^{2+}/Cu) = 0.34\ V$$

implies that for the reaction

$$\tfrac{1}{2}Cu^{2+}(aq) + \tfrac{1}{2}H_2(g) \rightleftharpoons \tfrac{1}{2}Cu(s) + H^+(aq)$$

the equilibrium constant, K_1, is given by

$$E^\circ(Cu^{2+}/Cu) = 0.34 = \frac{RT}{F}\ln K_1 \qquad (5.14)$$

A derivation of equation (5.14) from rigorous thermodynamic arguments may be found in chapter one, R. G. Compton and G. H. W. Sanders, "Electrode Potentials", (OCP 41).

where $K_1 = \dfrac{[H^+]}{[Cu^{2+}]^{\frac{1}{2}}p(H_2)^{\frac{1}{2}}} = 5.6 \times 10^5\ mol^{\frac{1}{2}}\ dm^{-\frac{3}{2}}\ atm^{-\frac{1}{2}}.$

Likewise, for Cu^+/Cu

$$Cu^+(aq) + \frac{1}{2}H_2(g) \rightleftharpoons Cu(s) + H^+(aq)$$

$$E^\circ(Cu^{2+}/Cu) = 0.52 = \frac{RT}{F}\ln K_2$$

where $K_2 = \dfrac{[H^+]}{[Cu^+]p(H_2)^{\frac{1}{2}}} = 6.2 \times 10^8\ atm^{-\frac{1}{2}}.$

It follows that for

$$2Cu^+(aq) \rightleftharpoons Cu(s) + Cu^{2+}(aq)$$

$$K = \frac{[Cu^{2+}]}{[Cu^+]^2} = \frac{K_2^2}{K_1^2}$$

$$= 1.2 \times 10^6\ mol^{-1}\ dm^3.$$

(b) For the reaction

$$2M^+(aq) \rightleftharpoons M^{2+}(aq) + M^+(aq),$$

the equilibrium constant,

$$K = \frac{[M^{2+}]}{[M^+]^2} = \frac{K_4^2}{K_3^2}$$

where $K_3 = \dfrac{[H^+]}{[M^{2+}]^{\frac{1}{2}}p(H_2)^{\frac{1}{2}}}$

and $K_4 = \dfrac{[H^+]}{[M^+]p(H_2)^{\frac{1}{2}}}.$

The equilibrium constants K_3 and K_4 relate to the equilibria

$$\tfrac{1}{2}M^{2+}(aq) + \tfrac{1}{2}H_2(g) \rightleftharpoons \tfrac{1}{2}M(s) + H^+(aq),$$

and

$$M^+(aq) + \tfrac{1}{2}H_2(g) \rightleftharpoons M(s) + H^+(aq)$$

so that

$$E^\circ(M^{2+}/M) = \frac{RT}{F}\ln K_3$$

and

$$E^\circ(M^+/M) = \frac{RT}{F}\ln K_4.$$

It follows that if

$$K > 1$$

then

$$K_4 > K_3$$

and

$$E^\circ(M^{2+}/M) > E^\circ(M^+/M)$$

if the disproportionation reaction is favourable.

6 Taking it further

6.1 Aims

In this chapter we shall briefly look at some particular examples that illustrate and extend the chemistry we have discussed in the previous chapters.

6.2 Drug design

Question 1

Fig. 6.1 shows the four base pairs involved in hydrogen bonding of double stranded DNA. The pairs thymine (T) and adenine (A), and cytosine (C) and guanine (G), are complementary.

Would there be any difference in the stability to heat of DNA sequences which contain a higher proportion of GC base pairs?

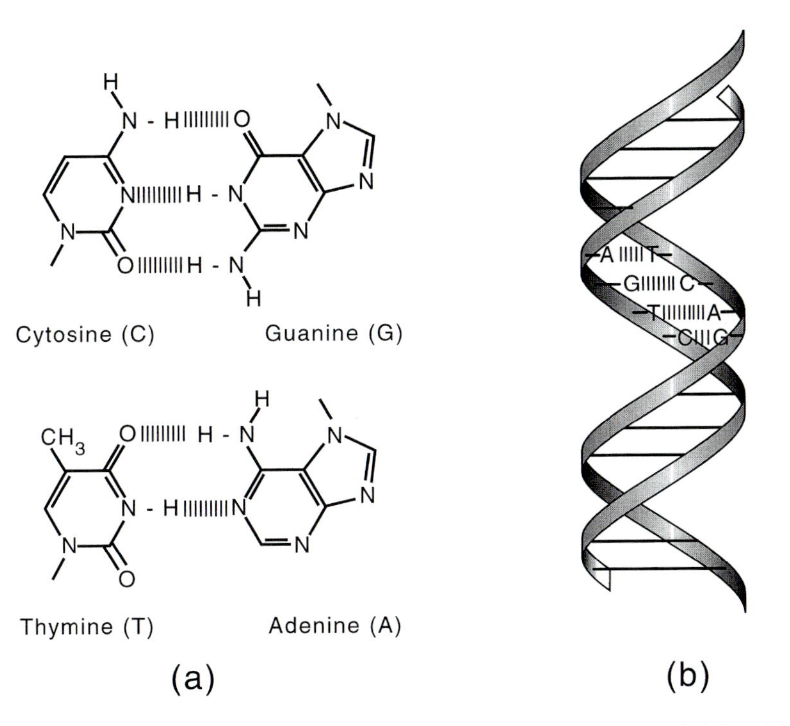

Fig. 6.1 (a) The complementary hydrogen bonding (ı ı ı ı) between the bases, C, G, T, and A which are the building blocks of DNA. (b) The DNA double helix is held together by hydrogen bonding between the two strands involving complementary base pairs (- A ı ı ı T -, - G ı ı ı C -).

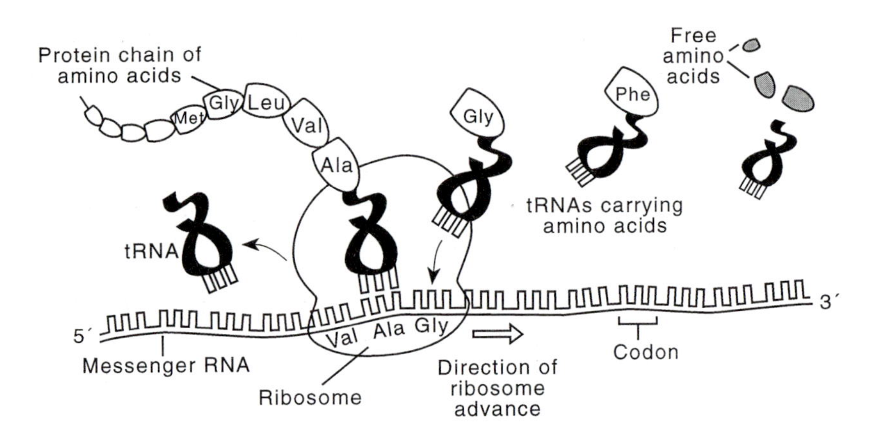

Fig. 6.2 The growth of an amino acid chain as it codes from the messenger RNA sequence.

Answer

The GC base pairs are held together by three hydrogen bonds whereas the AT pair are linked by two pairs. Therefore a higher proportion of GC base pairs might confer a greater stability towards heat.

Question 2

Fig. 6.2 shows the translation of RNA message into a protein. The DNA code contains four letters, G, C, A and T. The DNA codes for proteins which are made up of chains of amino acids of which there are twenty possibilities. Why does it take three DNA bases to code for each amino acid?

The use of computer modelling to predict the shape and behaviour of biologically active materials is described in G. H. Grant and W. G .Richards, "Computational Chemistry", (OCP 29).

The authors thank Professor W. Graham Richards and his research group for generously providing Figs. 6.1 and 6.2.

Answer

If one base coded for an amino acid only four labels would exist to distinguish twenty acids. Likewise if two labels were to be used a maximum of sixteen amino acids might be distinguish on a protein chain. It follows that a code of three bases is required to distinguish all twenty naturally occurring amino acids.

6.3 Michaelis-Menten kinetics

Question 3

An enzyme (E) catalyses the conversion of a substrate (S) into reaction products (P) according to the reaction scheme

$$E + S \underset{k_{-1}}{\overset{k_1}{\rightleftarrows}} ES$$

$$ES \xrightarrow{k_2} P + E$$

where ES denotes a bound substrate.

Enzyme kinetics and mechanism together with those of a host of other solution phase processes are described in B. G Cox, "Modern Liquid Phase Kinetics", (OCP 21).

(a) Use the steady-state approximation to derive an expression for the concentration of ES and hence show that the rate of formation of reaction product is

$$\frac{d[P]}{dt} = \frac{k_2[E]_0[S]}{K_M + [S]}$$

where K_M is a constant. $[E]_0$ and [S] refer respectively to the total concentration of enzyme and the concentration of unbound substrate in the system.

(b) Under what circumstances does the constant K_M approximate to the equilibrium constant describing the dissociation of ES?

(c) Measured values of d[P]/dt corresponding to particular values of [S] are as follows:

$10^6 \times [S]$ / mol dm^{-3}	5	10	20	50	100
$10^6 \times (d[P]/dt)$ / mol dm^{-3} s^{-1}	22	39	65	102	125

Calculate K_M.

(d) Ethanol is removed from blood through the action of the enzyme alcohol dehydrogenase in the liver. The loss has been found to follow zero order kinetics in both men and women with a rate constant varying between 0.05 to 0.08 mM min^{-1}. Explain and comment on the observation of zeroth order kinetics.

Answer

(a) Applying the steady-state approximation to ES,

$$\frac{d[ES]}{dt} = 0 = k_1[E][S] - (k_{-1} + k_2)[ES]$$

since the rate of making ES in the reaction of E with S must balance the sum of the rates at which ES is lost by the dissociation of ES and by conversion into products. It follows that the steady-state concentration of ES is

$$[ES] = \frac{k_1[E][S]}{(k_{-1} + k_2)}. \qquad (6.1)$$

However, if the total concentration of enzyme in the system is $[E]_0$,

$$[E]_0 = [E] + [ES]_{ss}. \qquad (6.2)$$

Therefore eliminating [E] from equations (6.1) and (6.2)

$$[ES]_{ss} = \frac{k_1[S]}{(k_{-1} + k_2)}([E]_0 - [ES]_{ss})$$

so that

$$[ES]_{ss}\left(1 + \frac{k_1[S]}{(k_{-1} + k_2)}\right) = \frac{k_1[E]_0[S]}{(k_{-1} + k_2)}$$

or

$$[ES]_{ss}\left(\frac{k_{-1} + k_2}{k_1} + [S]\right) = [S][E]_0.$$

Substituting

$$[ES]_{ss} = \frac{[E]_0[S]}{(K_M + [S])}$$

where the Michaelis constant,

$$K_M = \frac{k_{-1} + k_2}{k_1}$$

Finally,

$$\frac{d[P]}{dt} = \frac{k_2[E]_0[S]}{(K_M + [S])}. \qquad (6.3)$$

(b) If $k_2 << k_{-1}$ then

$$K_M = \frac{k_{-1} + k_2}{k_1} \approx \frac{k_{-1}}{k_1}$$

corresponding to the equilibrium constant describing the dissociation of ES.

(c) Equation (6.3) suggests that

$$\left(\frac{d[P]}{dt}\right)^{-1} = \frac{1}{k_2[E]_0} + \frac{K_M}{k_2[E]_0} \times \frac{1}{[S]}.$$

This implies that a plot of $\left(\frac{d[P]}{dt}\right)^{-1}$ against $[S]^{-1}$ will be linear if Michaelis-Menten kinetics apply; Fig. 6.3 shows the data in the question analysed in this way. The linear graph has

$$\text{slope} = 5.05\ s^{-1} = \frac{K_M}{k_2[E]_0},$$

$$\text{and intercept} = 2.95 \times 10^4\ mol^{-1}\ dm^3\ s^{-1} = \frac{1}{k_2[E]_0}.$$

Accordingly,

$$K_M = \frac{\text{slope}}{\text{intercept}} = 1.71 \times 10^{-4}\ mol\ dm^{-3}.$$

(d) If the removal of ethanol, S, by alcohol dehydrogenase, E, follows Michaelis-Menten kinetics then

$$\text{Rate} = \frac{k_2[E]_0[S]}{K_M + [S]}.$$

However if $[S] >> K_M$,

$$\text{Rate} = k_2[E]_0$$

and zeroth order kinetics are observed. It follows that if alcohol is removed in the manner observed,

$$k_2[E]_0 = 0.05 - 0.08\ mM\ min^{-1}$$

Note that it is the occurrence of zeroth order kinetics which permits the application of "rules of thumb" about alcohol metabolism specifying

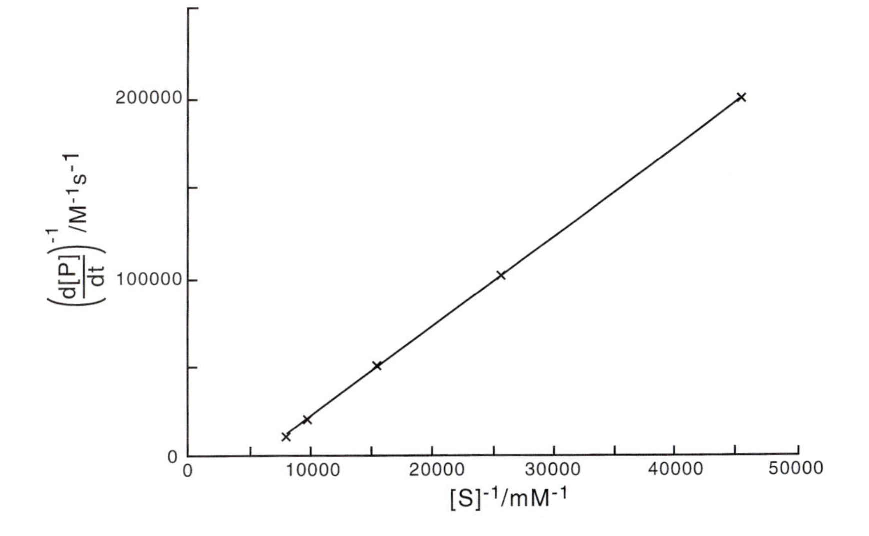

Fig. 6.3 Michaelis-Menten plot for the data in Question 1.

what volumes of beer are metabolised per hour. If the kinetics were other than zeroth order such generations would be inappropriate since the rate of removal would depend on how much alcohol was present.

Alcohol dehydrogenase contains two zinc atoms one of which provides an active site for the reaction

$$RCH_2OH + NAD^+ \rightarrow RCHO + NADH + H^+$$

where NAD^+ is the co-enzyme nicotinamide adenine dinucleotide. It has been shown that the latter is the first substrate to be bound to the enzyme and NADH is the last product to leave. The structure of the enzyme—NAD^+—alcohol complex has been deduced in the case of horse liver alcohol dehydrogenase and is shown in Fig. 6.4. The reaction mechanism involves

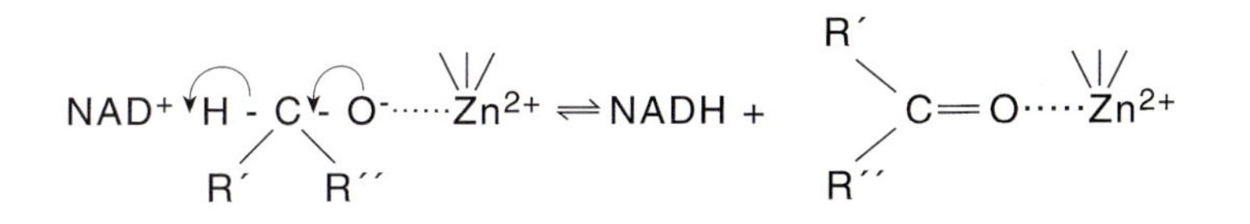

The role of metal ions in biology is described by D. E. Fenton, 'Biocoordination Chemistry', (OCP 25) and by P. C. Wilkins and R. G. Wilkins, "Inorganic Chemistry in Biology", (OCP 46). Alcohol clearance in humans has been reviewed by W. E. M. Lands, *Alcohol*, 15, (1998), 147.

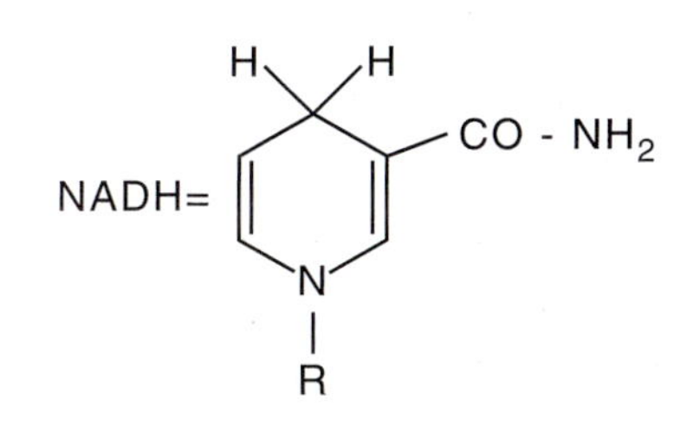

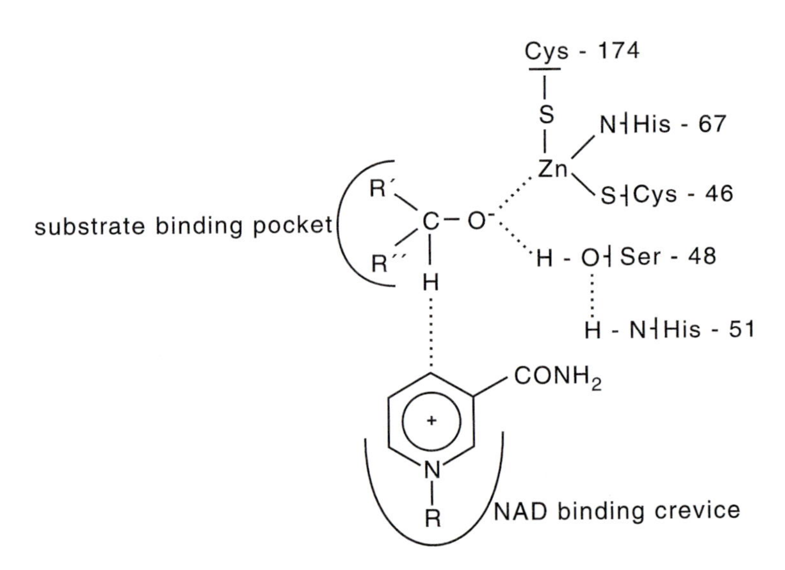

Fig. 6.4 The binding site of horse liver alcohol dehydrogenase: cys = cysteine, ser = serine, his = histidine

6.4 Electrolysis

Question 4

Fig. 6.5 shows a cyclic voltammogram recorded for the oxidation of N,N,N′,N′-tetramethylphenylenediamine, TMPD, (Fig. 6.6) in aqueous solu-

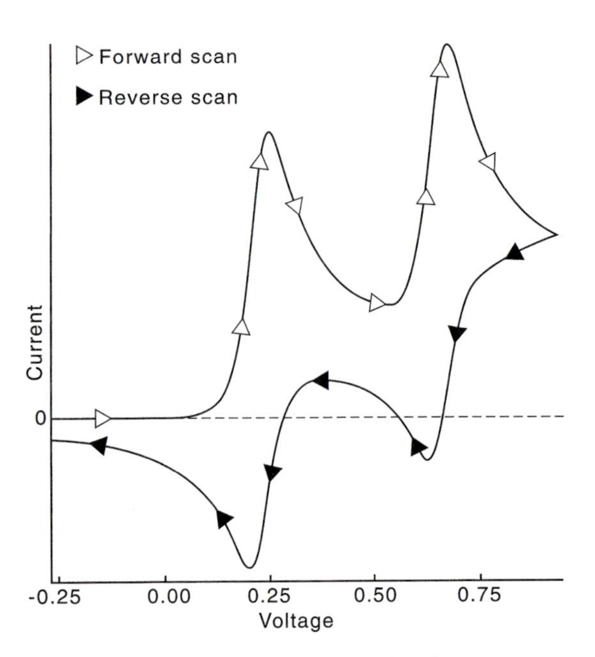

Fig. 6.5 A cyclic voltammogram recorded for the oxidation of N,N,N′,N′-tetramethylphenylenediamine with a scan rate of 500 mV s^{-1}.

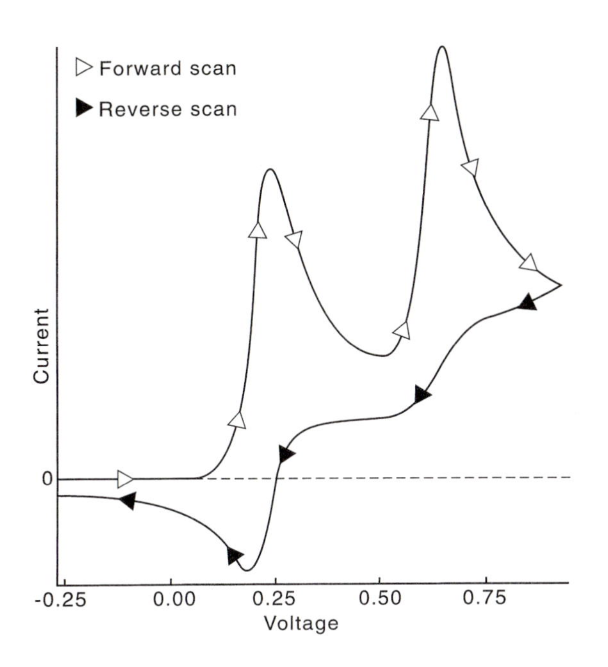

Fig. 6.7 A cyclic voltammogram recorded for the oxidation of N,N,N′,N′-tetramethylphenylenediamine with a scan rate of 10 mV s^{-1}.

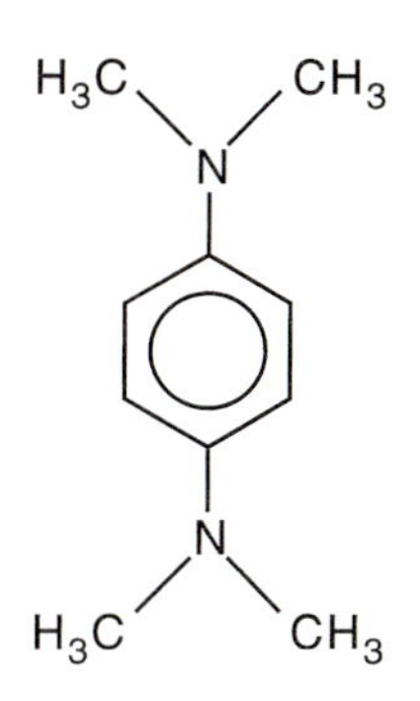

Fig. 6.6 The structure of N,N,N′,N′-tetramethylphenylenediamine.

tion at pH 7.0 using a voltage scan starting at −0.25 V and ending at 1.00 V. A scan rate of 500 mV s^{-1} was used. Explain the observed behaviour. Fig. 6.7 shows a corresponding voltammogram measured under identical conditions except that a scan rate of 10 mV s^{-1} was employed. Comment on the difference in behaviour from that found at the faster scan rates.

Electrolytic reactions are described by A. C. Fisher, "Electrode Dynamics", (OCP 34).

The data shown in Figs. 6.5 and 6.7 are from unpublished work by E. L. Beckett, R. G. Evans and P. J. Welford at Oxford University, 1998.

Answer

In the cyclic voltammetry experiment the voltage applied to an electrode is scanned in the manner shown in Fig. 6.8. The current flowing is plotted against voltage (Fig. 6.5 and Fig. 6.7) and for the case of interest two peaks are seen on the forward—positive going—scan. These represent the oxidations

$$\text{TMPD} - e^- \xrightarrow{0.25\text{ V}} \text{TMPD}^{+\bullet}$$

and

$$\text{TMPD}^{+\bullet} - e^- \xrightarrow{0.70\text{ V}} \text{TMPD}^{2+}$$

where the cation radical and the dication are shown in Fig. 6.9.

On the reverse scan, from 1.0 V to −0.25 V, the corresponding reduction takes place

$$\text{TMPD}^{2+} + e^- \longrightarrow \text{TMPD}^{+\bullet}$$
$$\text{TMPD}^{+\bullet} + e^- \longrightarrow \text{TMPD}$$

On slowing the voltage scan rate to 10 mV per second the peak corresponding to the reduction of TMPD^{2+} on the reverse scan is lost! This implies that the time taken to sweep the voltammogram is long in comparison with the lifetime of this reactive species. In fact TMPD^{2+} reacts with water with the displacement of dimethylamine.

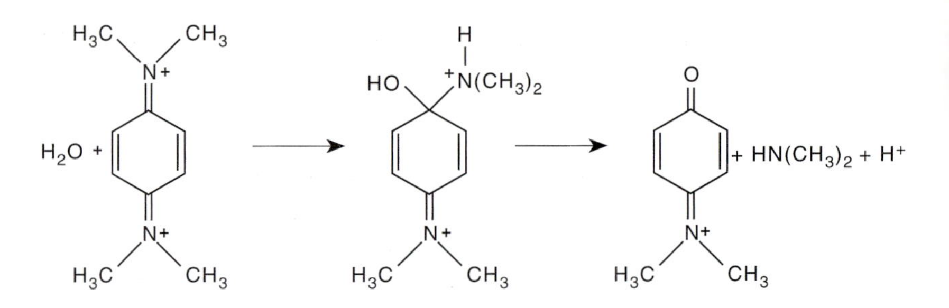

At a scan rate of 500 mV s^{-1} the time required for this reaction is very long in comparison with that needed to record the voltammogram and a reverse reduction of $TMPD^{2+}$ is seen. At 10 mV s^{-1} the opposite is the case and the peak corresponding to the reduction of $TMPD^{2+}$ is lost.

Cyclic voltammetry can be seen to provide an easy and elegant method to generate unstable and exotic species, and to study their chemistry.

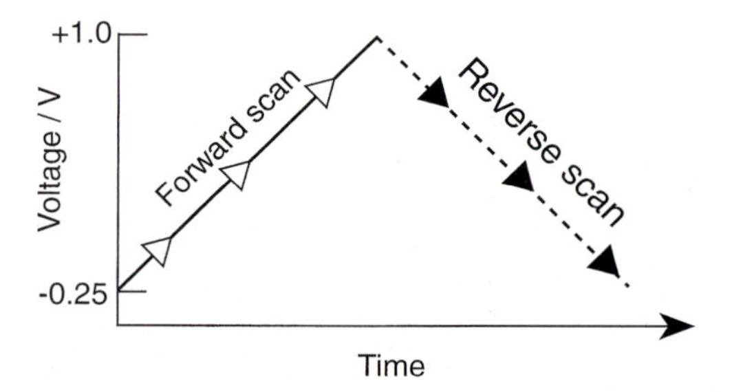

Fig. 6.8 The voltage applied to an electrode as a function of time in the cyclic voltammetry experiment.

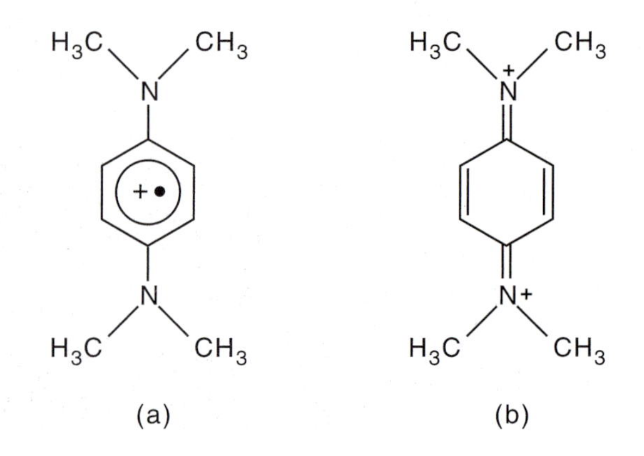

Fig. 6.9 The structures of (a) the cation radical $TMPD^{\bullet+}$ and (b) the dication $TMPD^{2+}$.

6.5 Looking at atoms: scanning tunnelling and atomic force microscopies

Question 5

In scanning tunnelling microscopy (STM) a sharp metallic tip is brought to within angström (Å) scale distances of a conducting surface and a small voltage applied between the tip and the sample: the resulting current is highly sensitive to their separation. If the tip is scanned over a selected area with the separation carefully maintained at a fixed value the "maps" of current provide atomically resolved images of the scanned surface. Whilst STM is clearly limited to samples having an appreciable conductivity, atomic force microscopy (AFM) can be applied to insulating samples. In this type of experiment the deflection (attraction or repulsion) of the tip is monitored as it is scanned over the surface (Fig. 6.10). Again atomic scale resolution can be achieved. Fig. 6.11 Shows an AFM image of the surface of a sodium chloride crystal. Calculate the Avogadro constant, N_A, given the crystal lattice spacing is 2.81 Å, the density of the NaCl crystal is 2.165 g cm^{-3} and the relative molecular mass (RMM) of NaCl is 58.5.

The STM and AFM techniques are described fully in G. A Attard and C. Barnes, "Surfaces", (OCP 59).

The authors thank Dr Drew Murray of TopoMetrix, Saffron Walden, Essex for kindly providing the image shown in Fig. 6.11.

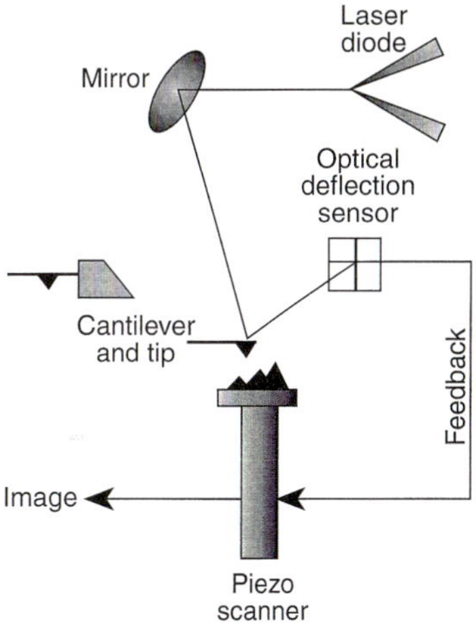

Laser diode
Mirror
Optical deflection sensor
Cantilever and tip
Feedback
Image
Piezo scanner

Fig. 6.10 A diagram of the experimental basis of AFM.

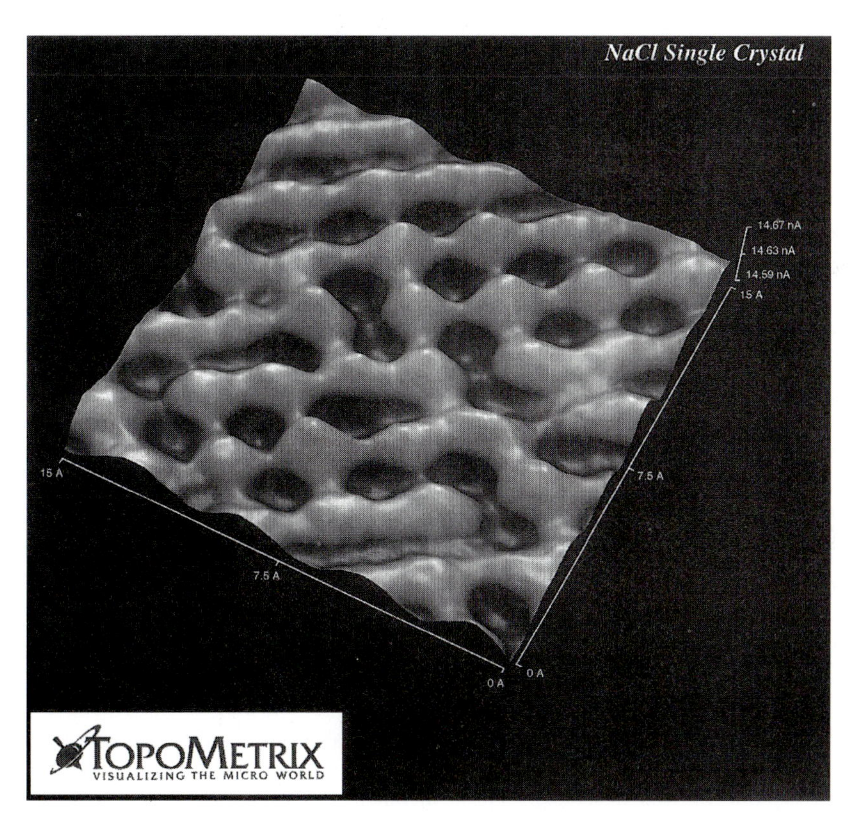

Fig. 6.11 An AFM image of the surface of NaCl.

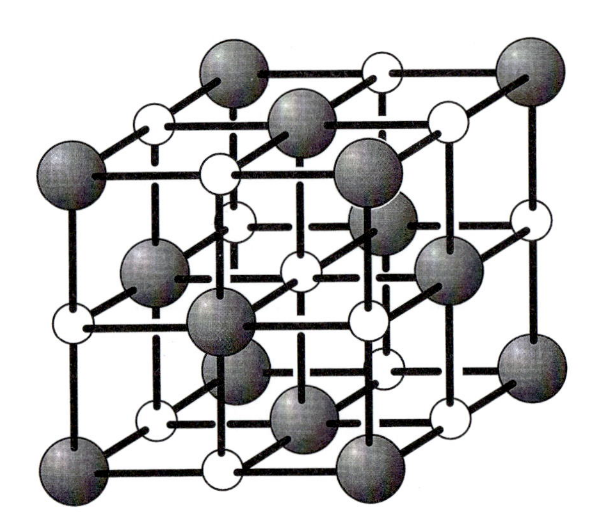

Fig. 6.12 The crystal lattice of NaCl.

Answer

Inspection of the AFM image shows that the surface structure is consistent with that expected for the NaCl crystal shown in Fig. 6.12 as inferred from X-ray crystallography.

Considering the sodium and chlorine atoms separately in Fig. 6.12 one can see how many of these are found in the unit cell shown by applying the following steps.

Step one: count how many atoms are in the body of the crystal. This is zero for Na and one for Cl.

Step two: count the number of atoms on each face of the unit cell remembering that only half the atom appears in this cell and half in the cell next to it. There are six Na atoms which means the unit cell contains three atoms in the single crystal. There are no Cl atoms on any face.

Step three: the atoms sharing an edge are shared amongst four unit cells. For the Na, no atoms are found on an edge. However the Cl has an atom on each of the crystal edges. Therefore twelve unit cells each donating a quarter of a Cl atom giving a total of three Cl atoms in the unit cell.

Step four. the atoms found at vertices are shared amongst eight unit cells. Na atoms are found on all of the vertices giving eight Na atoms each donating one eighth so the total number of Na in the unit cell drawn from the vertices is one. No Cl atoms are found at the vertices.

Step five: totalling up, its found that there are four Na atoms and four Cl atoms present and hence four ion pairs in the lattice fragment shown in Fig. 6.12. Each ion pair is separated by a distance of 2.81 Å. This means that the side of the cube drawn has a length 2×2.81 Å.

The mass of NaCl (RMM = 58.5) in the fragment is

$$\frac{4 \times 58.5}{N_A} = \frac{234}{N_A}$$

Since

$$\text{Density} = \frac{\text{Mass}}{\text{Volume}}$$

it follows that

$$2.165 = \frac{234}{N_A(2 \times 2.81 \times 10^{-8})^3}.$$

From this we can see that

$$N_A = \frac{234}{2.165 \times (2 \times 2.81 \times 10^{-8})^3}$$

$$N_A = 6.1 \times 10^{23}\ \text{mol}^{-1}.$$

This answer is agreeable close to the accepted value of 6.02×10^{23} mol^{-1}. However it is worth remarking that in carrying out the above calculation we have assumed that the ionic positions at the surface of the crystal are identical to these in the bulk crystal. This is by no means always the case and *surface reconstruction* is commonly encountered. In the case of NaCl it is thought that the chloride ions move out from the predicted ideal position and that the sodium ions move inwards from the latter. This arises since the larger negative ions are more polarisable than the smaller positive ions so that there is a larger induced dipole moment in the anion than in the cation. Accordingly the outer chloride ions move to increase the distance of the positive end of the dipole from the plane of the sodium ions and to decrease the distance of separation at the negative end. In practice relaxation occurs not only in the outer plane but in as many five planes before the structure of the ideal bulk crystal is attained. Fig. 6.13 illustrates the transition from the "surface" to the "bulk" behaviour.

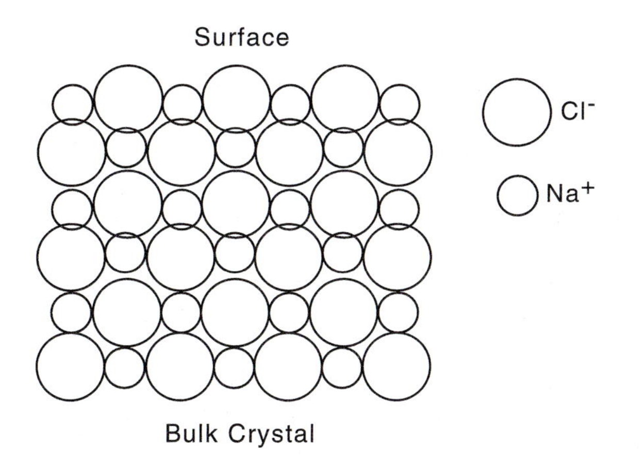

Fig. 6.13 The transition from "surface" to "bulk" behaviour in NaCl.

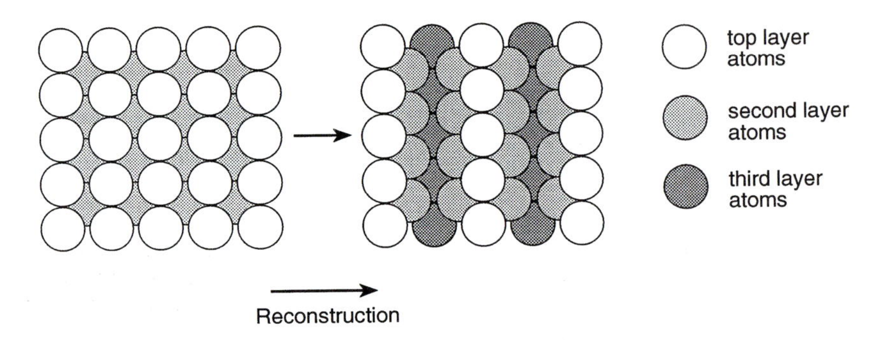

Fig 6.14 The reconstruction of a gold crystal face.

The case of the reconstruction of a silicon surface is covered in C. Lawrence, A. Rodger and R. Compton, "Foundations of Physical Chemistry", (OCP 40)

In other materials surface reconstruction can be much more dramatic. For example one surface of a gold single crystal can reconstruct so that there are 'missing rows' of atoms as seen in Fig. 6.14!

Index